Music, Geometry and Mathematics

The Source Code Revealed

By Derrick Scott van Heerden

Author of Mathemagical Music Production

Content copyright © 2018 Derrick Scott van Heerden. All rights reserved.

This publication contains material protected under International and Federal Copyright laws and treaties. Any unauthorized reprint or use of this material is prohibited.

No part of this book may be reproduced or transmitted in any form without written permission from the author, except in the manner of brief quotations embodied in critical articles and reviews. Please respect the author and the law and do not participate in or encourage piracy of copyrighted materials.

Contents	
Intro.	7
Dividing the octave:	9
Just intonation	11
Prime numbers	15
Regular numbers	19
Highly composite numbers	23
2, 3, and 5	27
Colors	29
Dividing the circle:	31
Cymatics	31
Sonic levitation	33
The geometry of just intonation	37
Irrational angles	45
Dividing the Sphere:	49
Platonic solids (Dice of the Gods)	49
Archimedean solids	59
Strength in objects	75
Ancient mathematics:	89
Mayan Long Count	89
Christians and 144000	91
The Yugas	93
Sumerian / Babylonian mathematics	95
Kings list	101
Numbers of the Gods:	103
Highest triad	105
Astral triad	109
7 most powerful deities	113
7 planetary deities	117
Other deities	121
In conclusion.	127

Intro

I decided to write this book to answer the one big question that everybody is asking, *why* do you find certain numbers like 144 and 432 in just intonation music scales, sacred geometry, the length of the day and the size of the Sun; various ancient texts and calendars, the measures in many ancient monuments, and so many other interesting places.

The answer was not really that hard to find. It all actually boiled down to just three numbers; the prime numbers 2, 3 and 5. They are, in fact, the mathematical "source code" behind all of these things.

5 limit music scales like Ptolemy's intense diatonic scale are entirely based on 2, 3 and 5. The Sumerians, Babylonians and various other ancient cultures also used mathematical systems based on 2, 3 and 5. They first divided the circle into 360 degrees and the day into 12 hours / 43200 seconds using such a system, which is why the same numbers show up in sacred geometry and measures of time.

Musically, 2, 3 and 5 represent the major chord, the chord that brings harmony and feelings of creative joy. The structure of this chord, however, also seems to provide strength in physical objects. Pyramids, geodesic domes, the Platonic and Archimedean solids for example are made by multiplying 2, 3 and 5 together in various combinations.

When I decided to look closer at the people behind all of this, the Sumerians and Babylonians, I found something really amazing. I found that in the religion of Sumer and Babylon, deities were assigned numbers that were all based on 2, 3 and 5. In this system the highest deities reveal the building blocks for music, sacred geometry and the ancient mathematical system behind it all. It is really the perfect synergy of mathematics, mythology and music when looked at in detail.

My suggestion is that you work through it with your keyboard or piano at hand. If you don't have one you can also download the enhanced eBook with audio, or look in the free intro of the standard eBook for links to my Youtube playlists. If you buy this paperback you will get an automatic discount on the eBooks.

You can use these to experience and, therefore, really understand the connections between all of these things. Just reading about stuff is not always the best way to understand and remember how it works. But when you actually *hear* geometric shapes, mathematical equations, and the positions of various deities in the heavens as musical tones and intervals, it all becomes crystal clear as you now have a feeling to attach to these things.

Dividing the octave

An octave is any frequency that's exactly double or half of itself. This is C to C, D to D, E to E etc. on a piano. The octave is considered the most harmonic interval after unison (the same note played over itself), which is why all C's or D's on a piano have the same name.

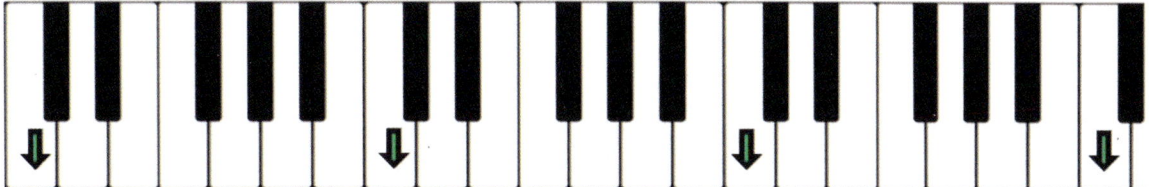

The 12 tone equal temperament scale that we use every day divides 1 octave into 12 *exactly-equal* parts. This is then repeated over more octaves to fill the keyboard.

Apart from the octave, which has a perfect 2/1 ratio in equal temperament, the other intervals in equal temp are based on and very close to pure ratio based intervals with the same names (major third, minor third, fifth etc.). The names just come from where the note is among the white keys starting with C; the third is the third note after C, the fourth is the fourth note, and the fifth is the fifth note after C. Obviously these intervals can be played in any key, it is just easier to understand them in C.

In scales with pure intervals, all notes are measured against the first note in the scale using a ratio instead of dividing the octave evenly. With these scales the notes are not exactly evenly spaced, and so they don't sound the same in any key like equal temperament does. So, in the end, equal temp was a tradeoff, losing the perfect harmony but gaining more musical options. This spawned a flood of creativity that changed the world of music forever.

In early times people only used these ratio based scales. They sounded better than equal temp in their root keys, but could not be played in *just any* key. Instruments tuned to these scales had to be re-tuned to make music in different keys, and were not suited for the mass market. Eventually people figured-out that dividing an octave into 12 equal parts would closely approximate the pure ratio based intervals that they were using quite well, and now you could play in any key. Other equal temperaments that divide the octave evenly into other amounts of parts did not sound so good or had too many notes for people to play easily, and so 12 was eventually settled-on as the best.

Just intonation

I have tried many ratio based scales, and have always found that this one sounds the best and is also the closest-sounding to equal temperament. Because the notes in this scale are so close to equal temp, you can use a normal piano or keyboard to get an idea of how they sound. You can look at this scale as the source scale; the intervals found here are the "original" most harmonious ones, by which the consonance of other versions of them (including the equal temp ones) can be measured.

Just intonation scale

C	1/1	Unison, perfect prime
C#	16/15	Minor diatonic semitone
D	9/8	Major whole tone
D#	6/5	Minor third
E	5/4	Major third
F	4/3	Perfect fourth
F#	7/5	Septimal tritone
G	3/2	Perfect fifth
G#	8/5	Minor sixth
A	5/3	Major sixth
A#	9/5	Just minor seventh
B	15/8	Classic major seventh
C	2/1	Octave

The way the ratios work mathematically is simple. You use them to calculate each frequency in relation to the first one (unison / perfect prime). For example; the 6/5 minor third is unison / perfect prime multiplied by 6 and divided by 5. It is the same for all ratios in the scale; just multiply perfect prime / unison by the first number in the ratio, and then divide it by the second number to get the frequency for that note.

Ratios with small whole numbers always sound better than ones with irrational or very large numbers. This shows you how simple mathematics and musical harmony really are closely connected, with mathematical simplicity always mirroring good sounding intervals.

The ratios also tell you between which harmonics each interval is found, in the harmonic series. For example; the 3/2 perfect fifth is found between harmonics 2 and 3, while the 5/4 major third is found between harmonics 4 and 5. This is a much simpler way to visualise ratio based intervals than to use mathematics.

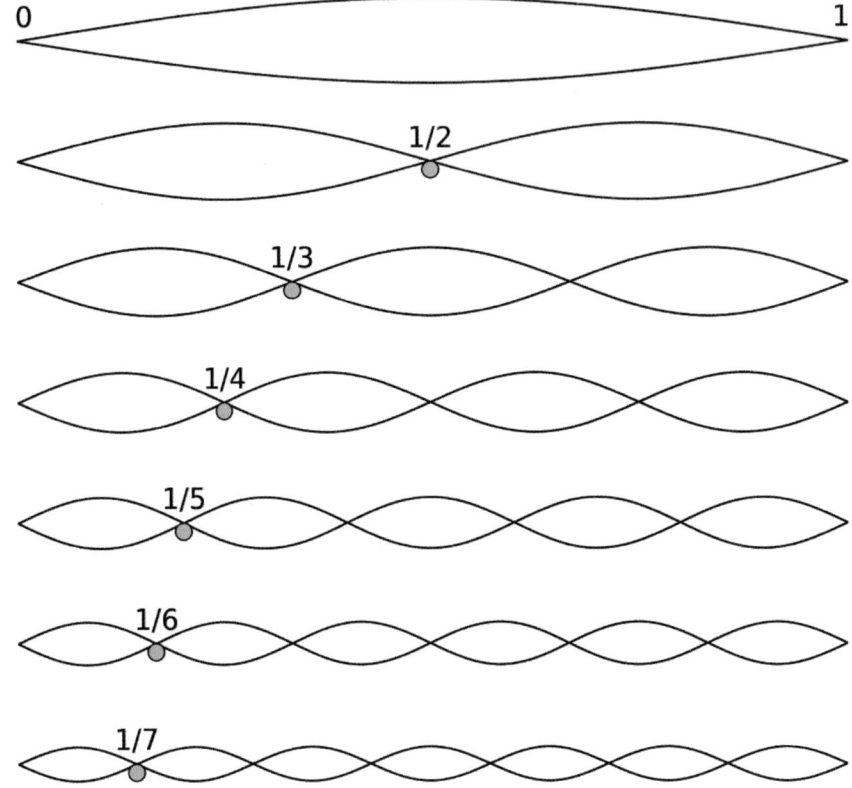

Better-sounding intervals are mostly found closer to the start of the harmonic series, which is why ratios with small numbers generally sound better. This is why the above scale can be considered *"the source with the best harmony"*, as it has the smallest possible whole number ratios for the 12 intervals.

Since most musical sounds are made from sine waves arranged according to the harmonic series, it makes sense that music scales made from these intervals would sound good when played with sounds that contain the same intervals in their harmonics. This is especially true with electronic music, as digital synthesizers follow the harmonic series precisely.

Spectrum analyzation showing the harmonic series in a tone:

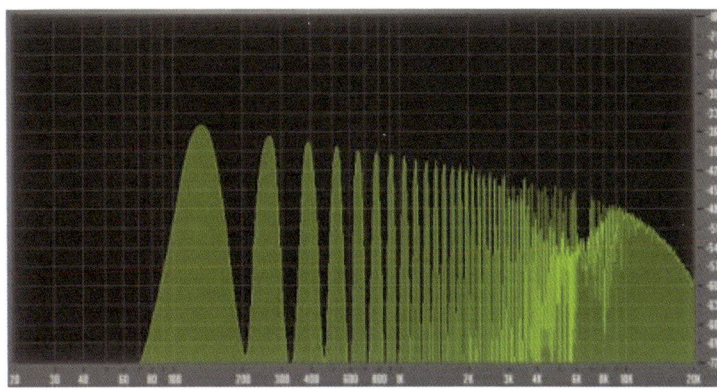

The same rule, where intervals found closer to the start of the series sound better, also applies to the harmonics in tones. Sounds that have louder harmonics below the 7th harmonic and softer ones above it sound nice and warm, while sounds with louder harmonics above the 7th harmonic tend to sound harsh and metallic.

If the whole numbers starting with 1 were Hz frequencies, they would make the harmonic series: 1, 2, 3, 4, 5, 6, 7, 8, 9, 10, 11, 12, 13, 14, 15, 16, 17, 18, 19, 20 etc. To bring this into hearing range as Hz frequencies you could just add some 0's: 100, 200, 300, 400, 500 etc. It would be the same thing, only in a higher key. So, when you do mathematics with whole numbers you are essentially doing *harmonic mathematics*.

You can get a feel for the first few harmonics on a normal piano; see how they get closer together as you go higher-up the series. Remember that the maths and vibration is the same. For example; 3 is 3 x 1, and 3 is also vibrating exactly 3 times as fast as 1. Just the same as 300 is 100 x 3, so 300 vibrates 3 times as fast as 100.

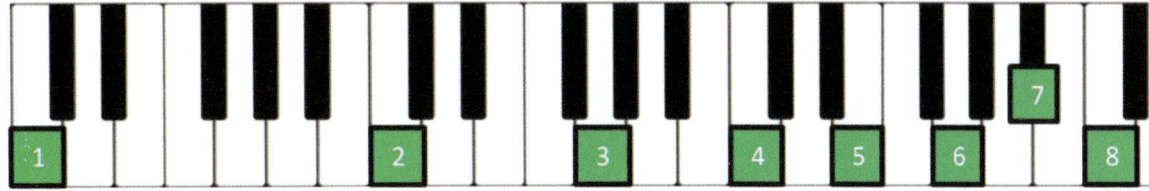

You can also use the chart above to see where ratio based intervals fit into the series (only ones with 8 or smaller numbers in their ratios will work with this chart). Just play the two keys that match the 2 numbers in that intervals ratio, to get an idea of how that interval sounds.

The ratios also tell you how many harmonics the two notes in that interval will share. For example; in the 3/2 perfect fifth, every third harmonic in the first note will be the same as every second harmonic in the second note. With the 5/4 major third, every fifth harmonic in the first note will be the same as every fourth harmonic in the second note. This partially explains why ratios with larger numbers tend to sound worse, as they have less matching harmonics.

It is not just as simple as this, though. Some small numbers sound worse in ratios than other larger ones. The 7/5 septimal tritone in the above just intonation scale for example, sounds pretty odd compared to some of the other intervals with larger ratios. The real 7/5 septimal tritone is close enough to the equal temp tritone, so-much-so that you can use it to get an idea of how this interval works in music. Even in equal temp tuning, the tritone is called the "devil's interval" because of this dissonance.

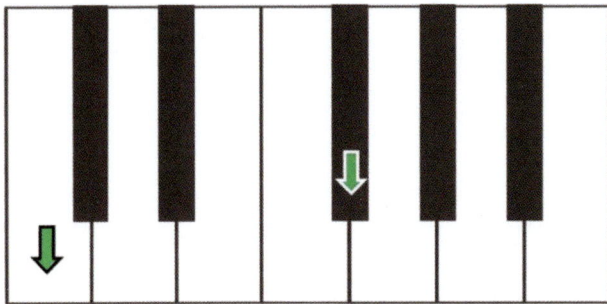

When you try this, you will find that whenever you use the tritone you will feel the urge to 'resolve' and move to another more pleasant interval. That is not to say that you must never use the tritone - sometimes it is *just* the thing to express more intense emotions.

Between harmonics 4 and 7 you will find another odd sounding interval called a harmonic seventh with a ratio of 7/4. It is not in the above just intonation scale and is quite a rough sounding interval that is quite a bit flatter than the same interval in equal temp. The musical context in which you would use both is about the same though.

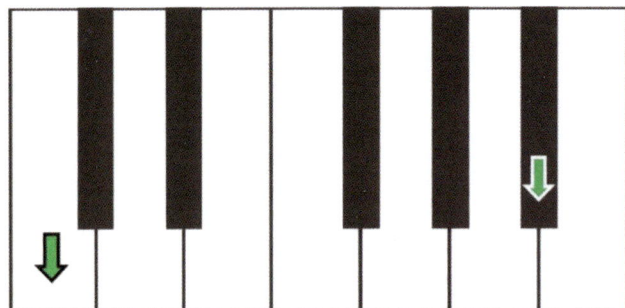

See how the odd sounding devil's interval and the harmonic seventh fall on a black keys, while the smoother ones are on white keys when you play in C. This is no accident; the piano was designed to work like this.

According to music theory, the 7/5 septimal tritone and the 7/4 harmonic seventh are different to all of the other intervals in the 12 tone just intonation scale. This is because in just intonation theory, ratio based intervals are defined by their "prime limit", and these two intervals have a different prime limit to the other intervals in the scale.

To better understand this we can look at only the white keys in the 12 tone just intonation scale. 7 notes are easier to study than 12, and they don't include the devil's Interval. This scale has been around for longer than the 12 tone version, and is called "Ptolemy's intense diatonic scale".

7 tone Ptolemy scale		
C	1/1	Unison, perfect prime
D	9/8	Major whole tone
E	5/4	Major third
F	4/3	Perfect fourth
G	3/2	Perfect fifth
A	5/3	Major sixth
B	15/8	Classic major seventh
C	2/1	Octave

To get an idea of how this scale sounds just play the white keys only, upwards on a piano starting on C. You will recognize it as the classic *do-re-mi-fa-so-la-ti* scale; the equal temperament C major diatonic scale. These are the keys of great harmony. Anybody can make good music using only these keys. Using all 12 keys can go bad easily if you don't know your music theory well.

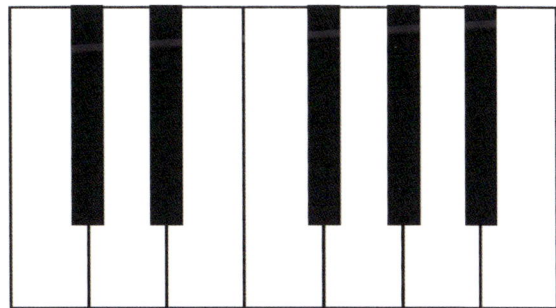

In music terms, Ptolemy's intense diatonic scale is called a 5-prime-limit tuning or a 5-limit scale. The intervals of 5-limit just intonation are made with ratios involving only products of the powers of 2, 3 or 5 (prime numbers limited to 5 or lower). The 12 tone version of the scale has the tritone with 7 in its ratio, so that scale is a 7-limit scale because of this one interval.

To understand this you need to know some basic math.

First you need to know what prime numbers are. A prime number is a natural number greater than 1 that has no positive divisors other than 1 and itself. For example; 5 is a prime number because it can only be divided by itself and 1. But 4 is not a prime number because it can also be divided by 2 (2 x 2 = 4). The first few prime numbers are 2, 3, 5, 7, 11, 13, 17, 19, 23, 29, 31, 37, 41, 43, 47, 53, 59, 61 etc.

Next you need to know what "products" of a number are. In mathematics, a "product" is the result of multiplying. For instance, 6 is a product of 2 and 3 (2 x 3 = 6).

Then you need to know that "powers" of a number are that number multiplied by itself. For example:

2 to the power of 1 = 2
2 to the power of 2 = 2 x 2 = 4
2 to the power of 3 = 2 x 2 x 2 = 8
2 to the power of 4 = 2 x 2 x 2 x 2 = 16

In written math, you'd just add a small number for powers, like this: '2^3' which represents 2 to the power of 3.

So, "products of powers of 2, 3 or 5" are basically 2, 3 and 5 and *any* numbers that can be made by multiplying (not adding) them together. If you look in the ratios in Ptolemy's intense diatonic scale, you will see that all of the numbers in them can be factored from the powers of 2, 3 and 5. For example; the prime factorization of 4 is 2 x 2. For 8 it is 2 × 2 × 2. For 9 it is 3 x 3, and for 15 it is 3 x 5.

If you try, you will see that you can never reach 7, 11, 13 or any other prime number larger than 5 by multiplying 2, 3 and 5 (prime numbers smaller than 5) together.

Powers of 2 represent octaves.
Powers of 3 represent perfect fifths (plus one octave).
Powers of 5 represent major thirds (plus two octaves).
Powers of 7 represent harmonic sevenths (plus 2 octaves)

So when you break this all down to its smallest parts, 5-limit tunings can be constructed entirely from octaves, fifths and major thirds. These 3 intervals represent a major chord and an octave, which explains why 5-limit intervals sound so nice, as they have a very harmonic source.

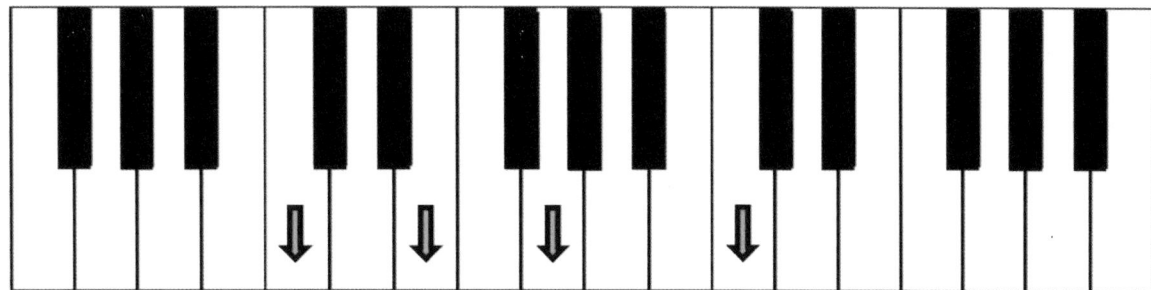

7-limit tunings on the other hand include the 7/4 harmonic seventh, These 4 intervals represent a major seventh chord and an octave. This chord is more jazzy and spicy, and not as smooth and relaxing as the plain major chord. This explains why 7-limit intervals sound odder. These are not necessarily bad - they are just intervals that may *want to resolve* to smoother sounding ones.

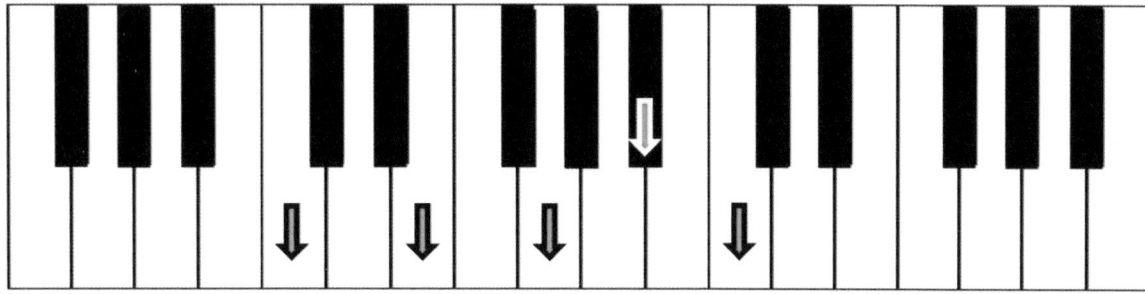

In Revelations (in the Bible), God rests on the seventh day... and does nothing. Maybe this is because nothing good could be made with 7? Perhaps *everything* was made with powers of 2, 3 and 5? Christians still consider Sunday, the seventh day in the week, to be a day of rest on which you should not work.

Regular numbers

A 5-limit scale can also be described as a scale with ratios containing only "regular numbers". This is because, in mathematics, regular numbers are numbers whose only prime divisors are 2, 3 and 5.

In number theory, regular numbers are also called 5-smooth numbers (you also get 7, 11, 13 or other prime-number-limited smooth numbers).

In computer science, regular numbers are often called Hamming numbers.

Regular numbers are also described as numbers that evenly divide powers of 60. In the study of Babylonian mathematics, these divisors of powers of 60 are called regular "sexagesimal" numbers. The Sumerians and Babylonians loved the properties of these numbers, and used them to make great advances in mathematics.

So, regular numbers, 5-smooth numbers, Hamming numbers and regular sexagesimal numbers are all the same thing. A set of very useful numbers made from products of powers of 2, 3 and 5.

The first few regular numbers are: 2, 3, 4, 5, 6, 8, 9, 10, 12, 15, 16, 18, 20, 24, 25, 27, 30, 32, 36, 40, 45, 48, 50, 54, 60, 64, 72, 75, 80, 81, 90, 96, 100, 108, 120, 125, 128, 135, 144, 150, 160, 162, 180, 192, 200, 216, 225, 240, 243, 250, 256, 270, 288, 300, 320, 324, 360…

These numbers are very musical if you think of them as Hz frequencies. In fact, you can find Ptolemy's intense diatonic scale and the 12-tone version previously discussed in this book just sitting among these numbers.

If, for example, you use 24 Hz as the fundamental frequency for Ptolemy's intense diatonic scale, you will find that the 7 frequencies are the nearly-consecutive regular numbers 24, 27, 30, 32, 36, 40, and 45. This is the same classic *do-re-mi-fa-so-la-ti* diatonic scale, except its played in G instead of C.

G major scale						
G	A	B	C	D	E	F#
24	27	30	32	36	40	45

When you play this scale in G, you need one black key (F#). In this context it is not the devil's interval, though. It is a nice sounding 15/8 classic major seventh, as it is measured against G and not C.

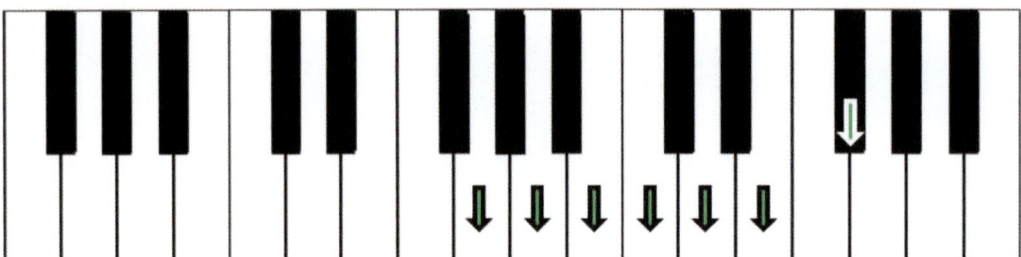

All higher octaves of these Hz frequencies are also regular numbers. This is because any powers of 2 (octaves) of a regular number also have to be regular numbers. This shows how the musical properties of tones and the mathematical properties of numbers stay very much the same when raised or lowered by octaves.

Ptolemy's intense diatonic scale over 5 octaves						
Unison	G	24	48	96	192	384
Whole tone	A	27	54	108	216	432
Major third	B	30	60	120	240	480
Fourth	C	32	64	128	256	512
Fifth	D	36	72	144	288	576
Major sixth	E	40	80	160	320	640
Major seventh	F#	45	90	180	360	720

Here you can see how octaves, fifths and major thirds / powers of 2, 3 and 5 can be used to get one octave these regular numbers. To go an octave higher, just add another x2.

Prime factorization				
2 x 2 x 2 x 2 x 2 x 2 x 3	$2^6 \times 3$	=	G	192
2 x 2 x 2 x 3 x 3 x 3	$2^3 \times 3^3$	=	A	216
2 x 2 x 2 x 2 x 3 x 5	$2^4 \times 3 \times 5$	=	B	240
2 x 2 x 2 x 2 x 2 x 2 x 2 x 2	2^8	=	C	256
2 x 2 x 2 x 2 x 2 x 3 x 3	$2^5 \times 3^2$	=	D	288
2 x 2 x 2 x 2 x 2 x 2 x 5	$2^6 \times 5$	=	E	320
2 x 2 x 2 x 3 x 3 x 5	$2^3 \times 3^2 \times 5$	=	F#	360

You can also find the 12 tone version of the scale (not including the devil's tritone) among the regular numbers, but to do this you must use 120 Hz as a reference pitch. When you do this the 7 frequencies that you get with 24 Hz as reference pitch are the same, only in a higher octave.

Here are the regular numbers up to 360 with the 12 tone version of the scale (not including the devil's tritone) in bold: 1, 2, 3, 4, 5, 6, 8, 9, 10, 12, 15, 16, 18, 20, 24, 25, 27, 30, 32, 36, 40, 45, 48, 50, 54, 60, 64, 72, 75, 80, 81, 90, 96, 100, 108, **120**, 125, **128**, **135**, **144**, **150**, **160**, 162, **180**, **192**, **200**, **216**, **225**, **240**, 243, 250, 256, 270, 288, 300, 320, 324, 360 etc. The devil's interval does not have a regular number because its ratio has 7 in it, and multiplying 120 Hz by 7 breaks the 5-limit rule.

12 tone just intonation			
Unison	1/1	B	120 Hz
Minor diatonic semitone	16/15	C	128 Hz
Major whole tone	9/8	C#	135 Hz
Minor third	6/5	D	144 Hz
Major third	5/4	D#	150 Hz
Perfect fourth	4/3	E	160 Hz
Septimal tritone	7/5	F	168 Hz
Perfect fifth	3/2	F#	180 Hz
Minor sixth	8/5	G	192 Hz
Major sixth	5/3	G#	200 Hz
Just minor seventh	9/5	A	216 Hz
Classic major seventh	15/8	A#	225 Hz
Octave	2/1	B	240 Hz

All of the regular numbers up to 80 are octaves of notes in this scale. After 80 some regular numbers are not in the scale, although they do still make good musical intervals. (the bold numbers are in the scale) **1, 2, 3, 4, 5, 6, 8, 9, 10, 12, 15, 16, 18, 20, 24, 25, 27, 30, 32, 36, 40, 45, 48, 50, 54, 60, 64, 72, 75, 80,** 81, **90, 96, 100, 108, 120,** 125, **128, 135, 144, 150, 160,** 162, **180, 192, 200, 216, 225, 240,** 243, 250, **256, 270, 288, 300, 320, 324, 360** etc.

As you can see, the connection between musical harmony and mathematics is very clear. The good sounding intervals represent the mathematically useful regular numbers, which in turn are based on the prime numbers up to 5. The odder sounding intervals, on the other hand, represent ratios that contain the prime number 7. This oddness extends to all prime numbers larger than 5, but I will explain that later in this book.

If you read my first book you will recognize 24 Hz and 120 Hz (in the previous chapter) as two of the best reference pitches to use to get whole numbers in all notes of the scale. Some other numbers like 36 and 360 were also very good. You may also remember that these are all highly composite numbers.

A highly composite number (or anti-prime) is a positive number that has more divisors than any other smaller positive number has. This is basically the opposite of a prime number which has no divisors other than 1 and itself. Obviously they work very well as reference pitches for whole number scales. You may have noticed that octaves of these are often used as reference pitches, for example you would usually use 192 Hz instead of 24 Hz. This is fine because the mathematical properties of numbers don't change much when they are adjusted over octaves. A higher octave might just have fewer divisors than another smaller number has.

The first few highly composite numbers are 1, 2, 4, 6, 12, 24, 36, 48, 60, 120, 180, 240, 360, 720, 840, 1260, 1680, 2520, 5040, 7560, 10080, 15120, 20160, 25200, 27720, 45360, 50400, 55440, 83160, 110880, 166320, 221760, 277200, 332640, 498960, 554400, 665280, 720720, 1081080, 1441440, 2162160 etc.

All of the highly composite numbers up to 720 are regular numbers. This is because highly composite numbers up to 720 only have 2, 3 and 5 as prime divisors. Above 720 you will find 7 as a prime divisor. Then, as you go higher you will find 11, 13 and so on (see chart below). So, highly composite numbers above 720 are not regular numbers.

12, 60, 120 and 360 which are super reference pitches for a low decimal just intonation scale, are not just highly composite numbers. They actually fall into a special category called "superior highly composite numbers". Simply put, these numbers have a stronger mathematical restriction and there are less of them.

The first few superior highly composite numbers are: 2, 6, 12, 60, 120, 360, 2520, 5040, 55440, 720720, 1441440, 4324320, 21621600 etc.

All superior highly composite numbers are also normal highly composite numbers. If you look at their prime divisors you can see how they get larger as you go higher up the sequence. I have only included a chart of superior highly composite numbers below, and not all of the highly composite numbers, because this chart is shorter and gets the point across easier. As you can see, going higher than 360 (or 720 with normal highly composite numbers) includes 7, 11 and 13 as prime factors. So the "best" of these are 2, 6, 12, 60, 120, 360, because they only include 2, 3 and 5 as prime factors.

Superior highly composite numbers	Prime divisors
2	2
6	$2 \cdot 3$
12	$2^2 \cdot 3$
60	$2^2 \cdot 3 \cdot 5$
120	$2^3 \cdot 3 \cdot 5$
360	$2^3 \cdot 3^2 \cdot 5$
2520	$2^3 \cdot 3^2 \cdot 5 \cdot 7$
5040	$2^4 \cdot 3^2 \cdot 5 \cdot 7$
55440	$2^4 \cdot 3^2 \cdot 5 \cdot 7 \cdot 11$
720720	$2^4 \cdot 3^2 \cdot 5 \cdot 7 \cdot 11 \cdot 13$

These superior highly composite numbers, especially 12, 60, and 360, have been known for their amazing mathematical properties for thousands of years. This is mainly because they have so many divisors, and are therefore very useful for dividing things up into equal parts.

This is why we have 360 degrees in a circle and 12 inches in a foot. And also why we have 60 seconds in a minute, and 60 minutes in an hour; 12 hours in a day, 12 months in a year / 12 Zodiacal signs, and the approximately 25920 year "great year" which is divided into 12 cycles of about 2160 years each.

In case you didn't notice, dividing a circle into 12 parts is exactly the same as dividing an octave into 12 parts.

The Sumerians, Babylonians, Mayans and a few other ancient cultures had 360 day per year calendars instead of the true 365. It seems like some of them really thought there were 360 days, while others knew that there were 365, but just ignored the extra days. They did this because 360 is a superior highly composite number, and it has many divisors which are all regular numbers. This made it easy to divide time into even pieces. 365 is not a highly composite or regular number, and only has 1, 5, 73 and 365 as divisors. 365 is actually called a "semi-prime" because it has so few divisors.

The divisors of 360 are: 1, 2, 3, 4, 5, 6, 8, 9, 10, 12, 15, 18, 20, 24, 30, 36, 40, 45, 60, 72, 90, 120, 180 and 360 - exactly 24 divisors. All divisors of 360 are regular numbers, and all divisors of 12 and 60 are also divisors of 360.

Divisors of 12, 60 and 360																							
12	1	2	3	4		6			12														
60		2	3	4	5	6		10	12	15		20		30			60						
360	1	2	3	4	5	6	8	9	10	12	15	18	20	24	30	36	40	45	60	72	90	120	180	360

All divisors of 360 are octaves of notes in Ptolemy's intense diatonic scale, so the divisors of 12, 60, and 360 are a very musical set of numbers. Here is one octave of the divisors of 360, the rest are just more octaves of these notes. As you can see, only one note (A = 27) is missing, otherwise it is the full Ptolemy's intense diatonic scale in G = 24 Hz.

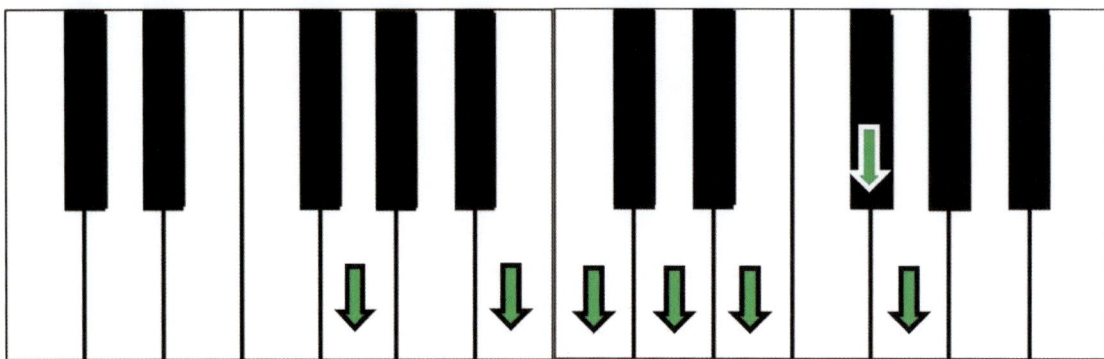

It is interesting that numbers which can divide 12, 60 and 360 evenly mathematically represent such good sounding intervals, while 7 (and higher prime number) based intervals like the Devil's tritone can't divide 360 evenly and also sound odd.

2, 3, AND 5

So, the regular numbers, highly composite and superior highly composite numbers, and all of this *mathematical and musical goodness* come from products of powers of the prime numbers 2, 3, and 5.

If you play 2, 3, and 5 on a piano, you will hear a type of inverted major chord. You just need to change the octaves of the notes to play all other types of major triad chords. Since octaves = powers of 2, and shifting in powers of 2 is allowed in 5-limit tuning, you can shift the notes over octaves as much as you want without breaking the 5-limit rule.

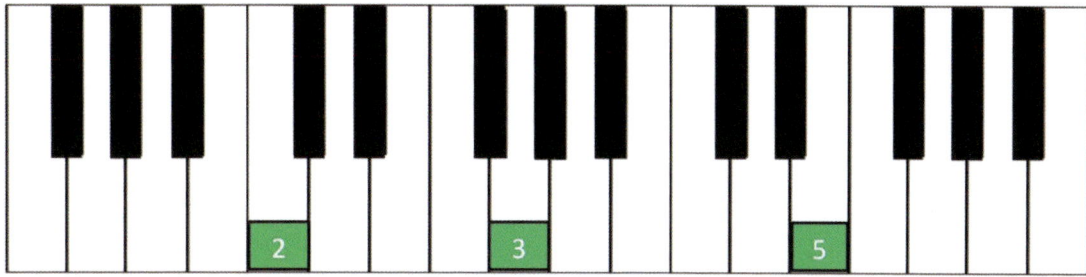

I personally find that the way the exact 2-3-5 chord spreads harmonics by having wider gaps between the notes sounds really good. I find it even more pleasing than the "normal" 4-5-6 major chord which compresses more harmonics together. Below you can use the harmonic series to see how to play a 2-3-5 or 3-4-5 or 4-5-6 C major chord by changing the octaves of the notes.

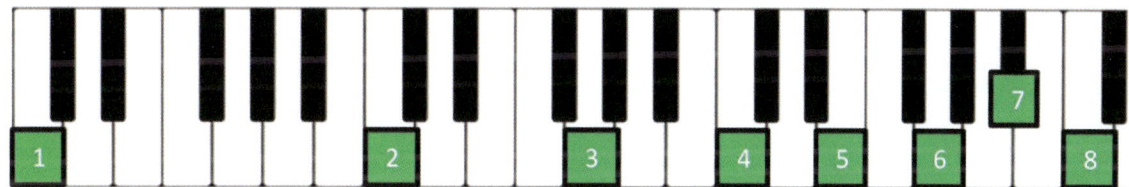

The fact that there are so many major chords near the start of the harmonic series explains why major chords sound so good when they are played on instruments with harmonics arranged according to the same harmonic series.

Color fits into all of this in a really interesting way. If you take the Hz frequencies for Ptolemy's intense diatonic scale, with 192 Hz (Octave of 24 Hz) as reference pitch, and raise them by 40 octaves, they will now be light frequencies. As Hz frequencies, these numbers are very, very long. And so light is measured in nanometers, which is the *length* of the light wave and not its frequency.

So, if you raise the scale by 40 octaves and convert the long Hz frequencies to nanometers to see their "true colors", some amazing things are revealed.

Tones and colors

Note	Hz	Nanometers	Color
G	192 Hz	710.0500 nm	Red
A	216 Hz	631.1600 nm	Orange
B	240 Hz	568.0400 nm	Yellow
C	256 Hz	532.5400 nm	Green
D	288 Hz	473.3700 nm	Blue
E	320 Hz	426.0300 nm	Purple
F#	360 Hz	378.6900 nm	Indigo

Firstly, the three notes in the G major chord (G – B – D) each fall into one of the three color bands occupied by the three primary colors red, yellow and blue.

All 7 notes in the scale fall into the color bands for the 7 colors in the rainbow. These are the 7 colors that are really unique and have unique names. If you use the 12 tone scale the rest of the notes would just be shades of these colors, like dark red or light blue etc. Some of these shades do have names, but to me they are just lighter or darker variations of the main 7 colors. Because each color covers a band and not a single frequency, any 12 tone chromatic scale - even equal temp tuned to 440 Hz - will still work perfectly for this.

If you use complementary colors instead of the actual colors, then C becomes red and the C major chord represents the 3 primary colors.

C	Red	Unison, perfect prime
C#		Minor diatonic semitone
D	Orange	Major whole tone
D#		Minor third
E	Yellow	Major third
F	Green	Perfect fourth
F#		Septimal tritone
G	Blue	Perfect fifth
G#		Minor sixth
A	Indigo	Major sixth
A#		Just minor seventh
B	Violet	Classic major seventh
C	Red	Octave

This is all very interesting because with paint as the medium the three primary colors are used to mix the 7 colors. These 7 are then used to mix the 12 chromatic colors. This is the same as the way the major chord triad is used to make the 7-tone diatonic scale, which is then used to make the 12-tone chromatic scale. This major chord which represents the three primary colors also represents the prime numbers 2, 3 and 5.

This mirrors the Kabbalistic text Sepher Yetzirah ("The Book of Formation") which breaks the 22 letters of the Hebrew alphabet up into three sets: 3 mother Letters, 7 double Letters and 12 elemental Letters.

I found a similar story in the Bible with 3 representing the body, soul and spirit, or the father, the son and the Holy Ghost for the Catholics. 7 represents the number of days in the first week of creation, and is therefore the number of completion or perfection, and 12 represents the 12 disciples or the 12 tribes of Israel.

It also mirrors the Babylonian system (later in this book) quite well. That system has: 3 highest deities, 7 most powerful deities and 12 "places" for younger deities.

The basic picture here seems to be that 3 creates 7, which then creates 12.

This is a good way to visualize music, with the closest music to "source" being just the 3 tone major chord (droning meditation music). The next closest is music made with the 7 tone diatonic scale, and the farthest - which includes the devil's interval - being music made with the 12 tone chromatic scale.

Dividing the Circle

You likely don't yet know this, but the intervals in any ratio based scale can be represented by two shapes. A 4/3 fourth represents a square and a triangle. A 5/4 major third represents a pentagon and a square – and so on. Each number in the any ratio represents a regular polygon with the same amount of sides as that number.

Cymatics

This may sound like my own stupid idea, but physical proof of this exists. To understand it you need to look at cymatics. This is the art of vibrating water or other particles on a vibrating plate. While a lot can be learned from this, the results are always dependent on the size and shape of the plate. If you watch videos of this you will see that as you slowly raise the frequency, certain ones make amazing patterns while others don't. All that is happening is that the frequency is hitting the resonant frequency of the plate and its harmonics. When a harmonic is reached the plate vibrates in perfect multiples of its size, and geometric patterns emerge. These patterns become more complex with higher tones, because higher harmonics divide the plate into smaller portions.

This is easy to demonstrate with a length of rope about 2 or 3-meters long. Just tie one end to something solid, and then start to move the other end up and down slowly. Eventually you will find a "sweet" spot; a speed that can make a perfect arch (as seen in the top part of the image below) with very little effort on your part. If you now move it exactly twice as fast it will go messy, and then suddenly make a double arch (as seen in the second part of the image). It will alternate up on the left and down on the right, and then swop over for each wave. If you now move it 3 times as fast, you will get 3 arches. These will alternate with the middle up and the sides down and then swop. You can go on to 4 and maybe 5, and then at that speed... your hand will get tired! What you are doing here is dividing the rope by the harmonic ratios 1/1 –1/2 - 1/3 etc.

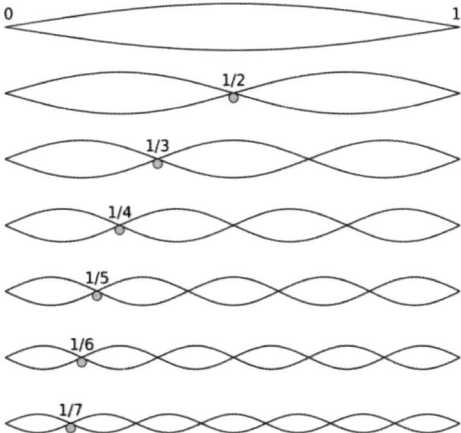

To get closer to the source of what is happening, you can use sonic levitation. Then it becomes possible to use sound to levitate and vibrate a drop of water, but with no plate at all. This works by creating a standing wave with stable nodes that can hold small objects. Now you can see the "source shape" for each harmonic, upon which all of the more complex ones on the plate are based.

There is a video you should see before reading further called "Shape oscillation of a levitated drop in an acoustic field". You can find it on Youtube.

In it they do something amazing. They play the fundamental frequency of the levitated drop of water, and it flattens into a disk (circle). Then they play the second harmonic (octave) of this pitch, and it becomes elongated like an oval. This oval, however, is also alternating / oscillating just like the rope mentioned before does. The ends move together and the sides move apart, forming new ends so that it's like a fat "plus symbol", but with only one oval visible at a time, at the peak of each oscillation. When they play the third harmonic the water forms a triangle, also oscillating with another one so that both together would be a Star of David, but only one triangle was fully formed at a time.

And so they go on. Each harmonic's shape has the same amount of sides as its sequence number: 1 = circle, 2 = sausage, 3 = triangle, 4 = square, 5 = pentagon, and so on. This water drop is actually behaving exactly the same as the swinging rope, only the "rope" is circular and is made from the surface tension of the water. You also find other videos showing star shapes in vibrating bowls and such things (although these are never as accurate as a levitated drop). So, no doubt can exist that the harmonic series is directly connected to the polygons.

If you join the corners of these stars with straight lines, you get regular polygons. Each harmonic of the resonant frequency of the drop of water = a polygon with the same amount of sides as that harmonic's number:

1 = circle (one side)
2 = oval (two sides)
3 = triangle (three sides)
4 = square (four sides)
5 = pentagon
6 = hexagon
7 = heptagon
8 = octagon
9 = nonagon
10 = decagon
11 = hendecagon
12 = dodecagon
13 = tridecagon
14 = tetradecagon
15 = pentadecagon
16 = hexadecagon
17 = Heptadecagon
18 = Octadecagon
19 = Enneadecagon
20 = Icosagon

You can hear how the first few of these shapes relate to each other and how represent musical intervals using a piano to play the start of the harmonic series and looking at the following image:

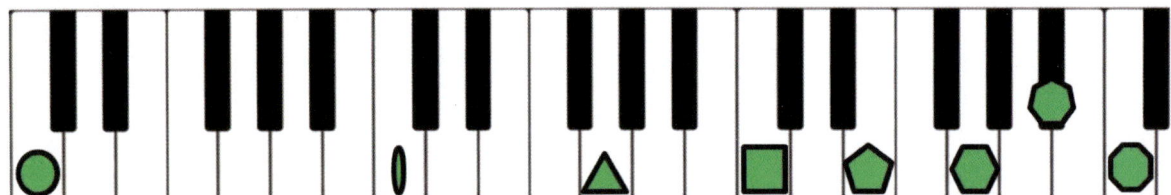

Higher octaves of each polygon contain the same polygon x 2. This shows you how the octave is the most harmonic interval using geometry, as it is the only interval that doubles the same polygon inside the new one.

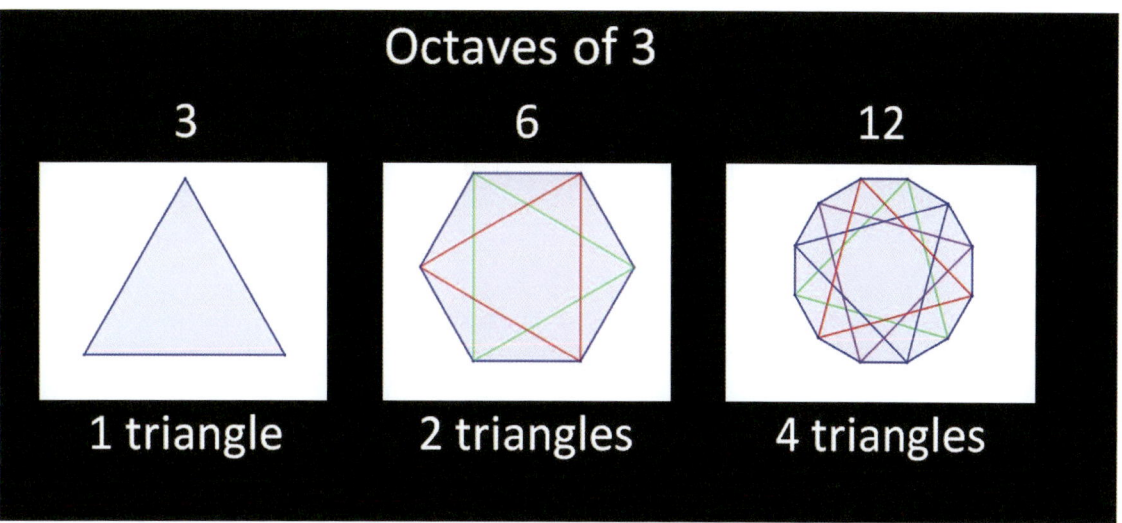

Remember that in the levitated drop of water, the square actually contains two ovals. So while in geometry the triangle is considered to be the smallest polygon, it vibration it is really the oval. Although it does not look like it at first, the oval does also equally divide a circle and its sides are all equal in shape and length, so it can also be a polygon. In the next chapter I will represent the two ovals in a square as two straight lines (for example 4 in the image below). I will not add the lines when a polygon contains multiple squares, so in those you will just have to imagine them (for example 8 or 16 in the image below).

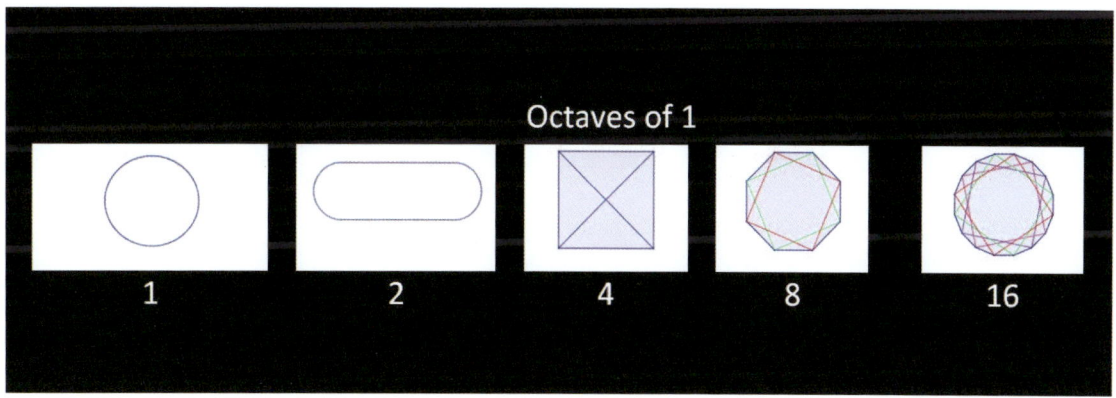

Since musical ratios also tell you which 2 harmonics that interval is found between in the harmonic series, musical ratios can also be seen as two shapes instead of two numbers.

Here are the 7 and 12-tone versions of Ptolemy's intense diatonic scale as polygons and numbers. I have inserted smaller polygons inside the polygons that *can* fit smaller ones inside themselves. With them you can visualize the prime factors of the larger polygons as shapes. This is a great way to learn about prime factors and would be good at schools…

Remember that I did not add ovals or lines in the squares when there are more than one of them in a polygon.

15 can fit 3 pentagons or 5 triangles in itself, so there are two images for 15.

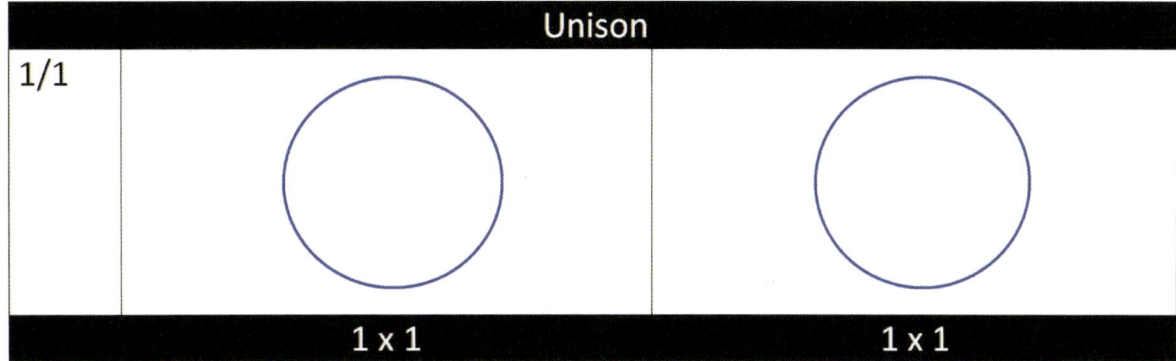

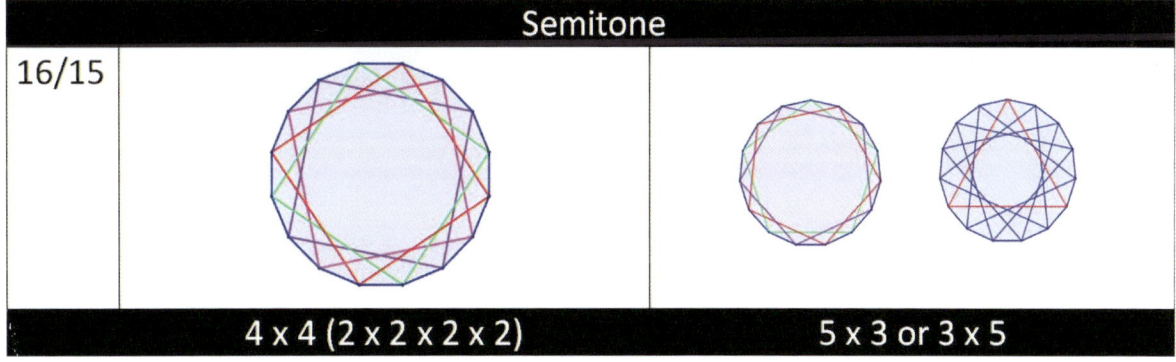

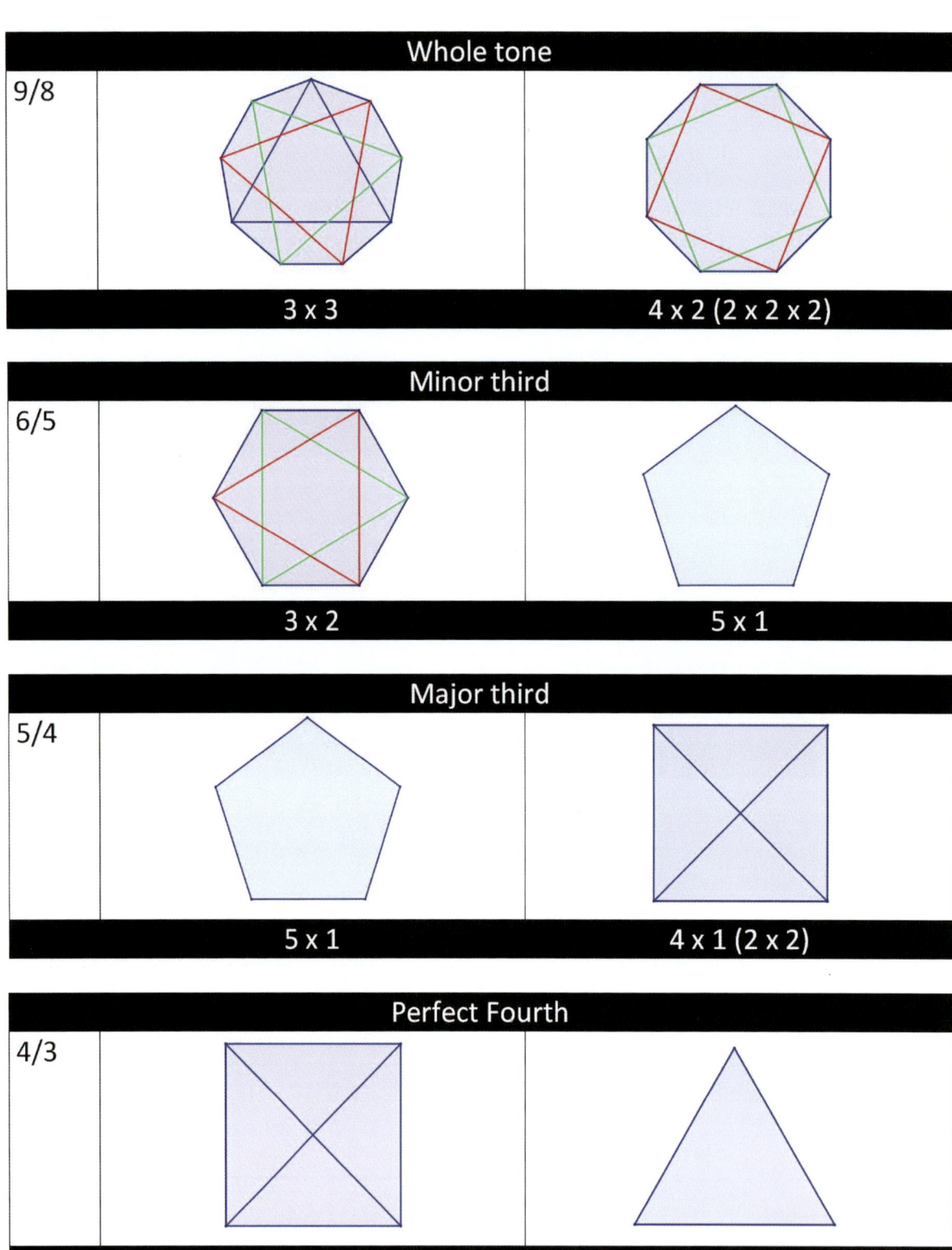

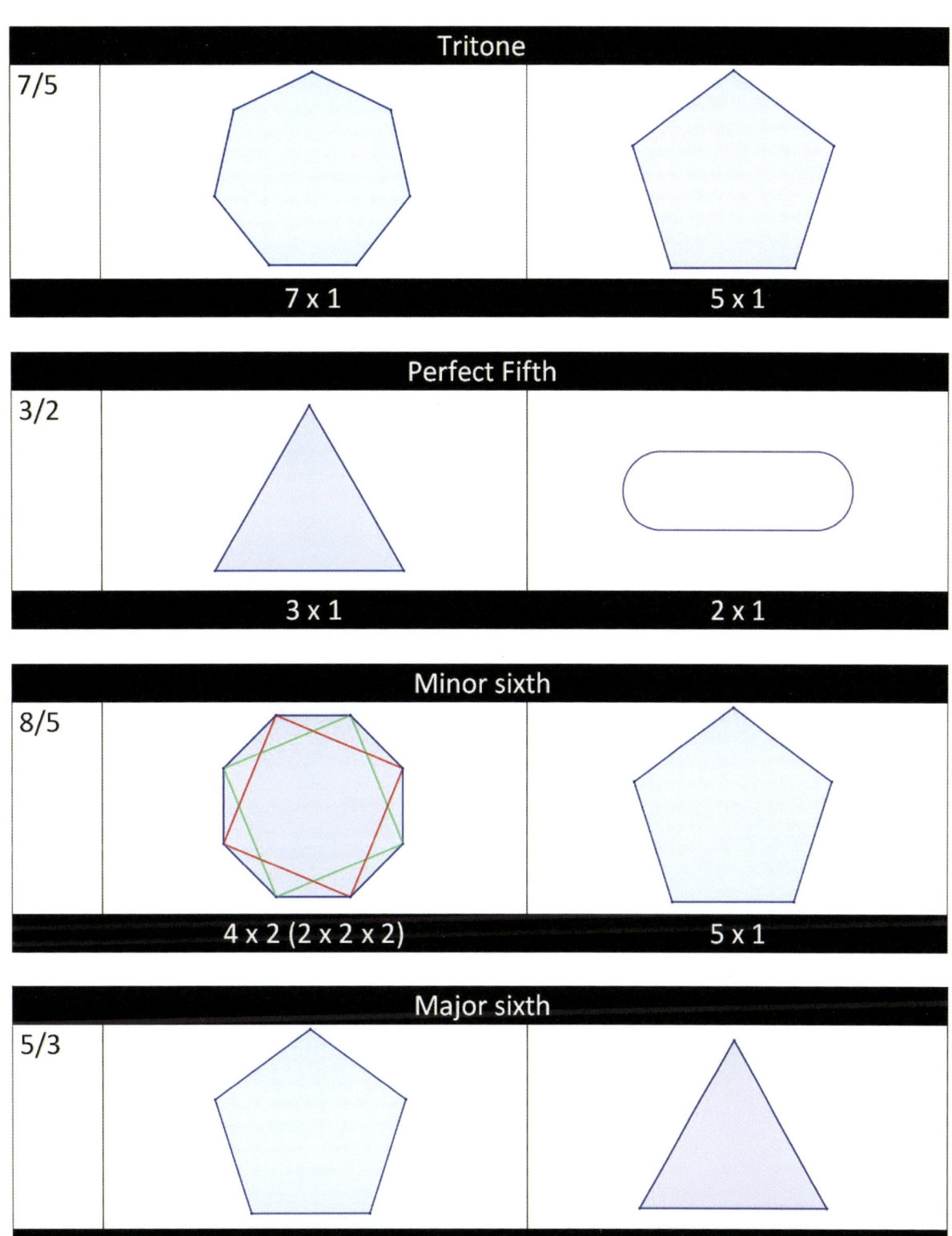

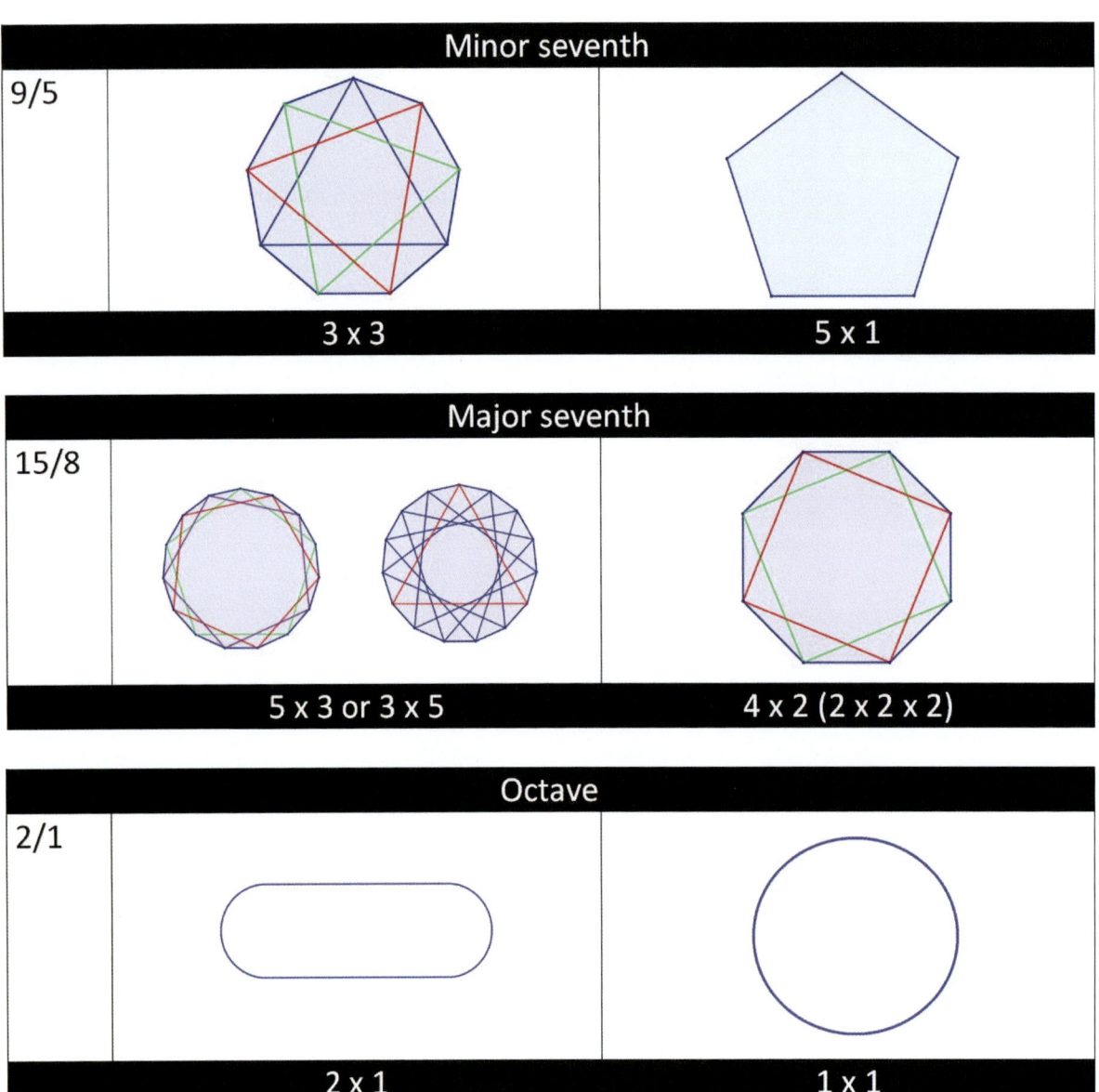

See how all of the 5-limit intervals (and not the 7-limit tritone) are made from ovals, triangles, pentagons, or larger polygons that can fit regular number amounts of them inside themselves. Unison and the octave contain circles, but the circle is the root of the scale and not an interval.

So, all of the intervals in any 5-limit scale have the "geometric prime factors" of 2, 3, and 5-sided polygons (ovals, triangles and pentagons). This is not really surprising, because these are 5-limit intervals based on the prime numbers 2, 3, and 5.

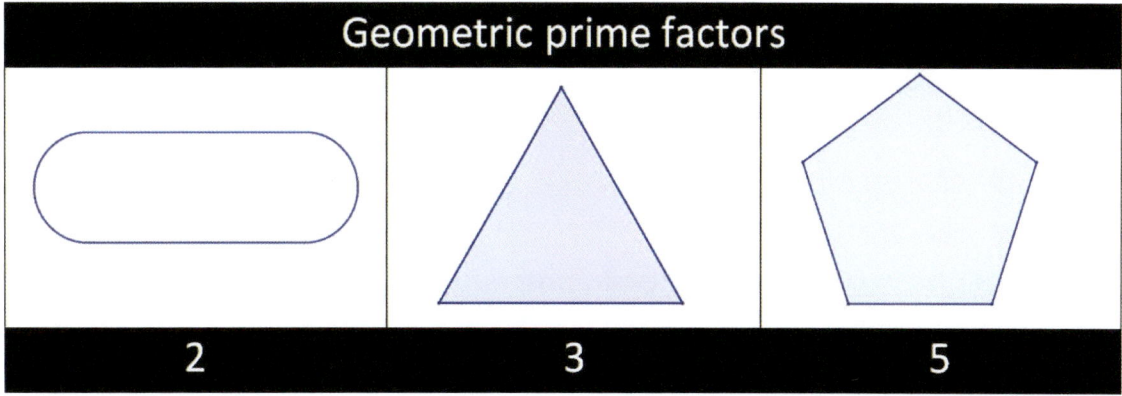

On the next page you can see some more intervals that are used in some micro-tuned music, but that are not very close to intervals in equal temp. It is quite interesting how dividing an octave into 12 equal parts (equal temp) avoids strange intervals while mirroring the 5 limit ones quite well. If you have read my previous books you will find all of these intervals in the "Dwarf scale".

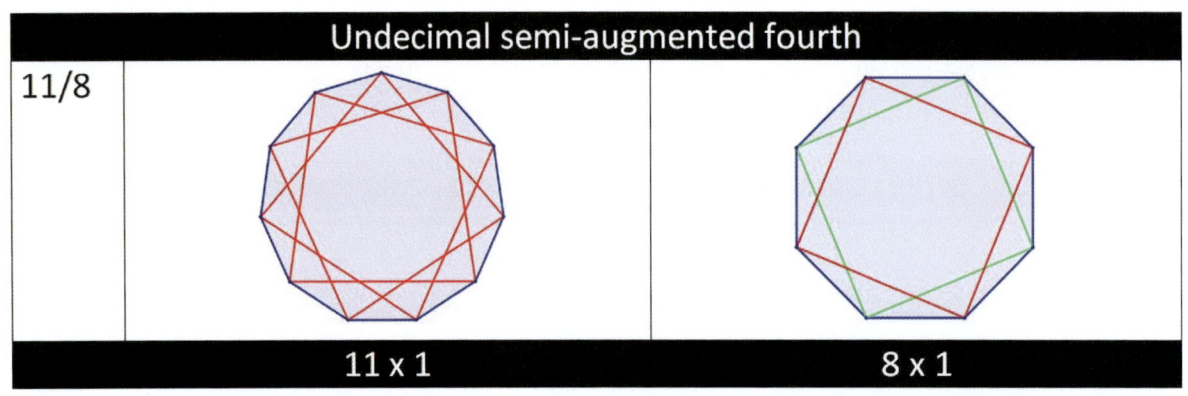
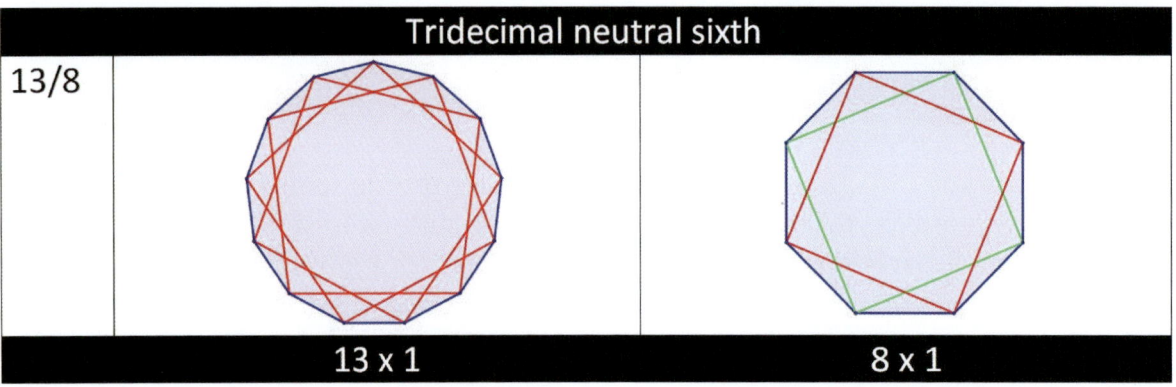
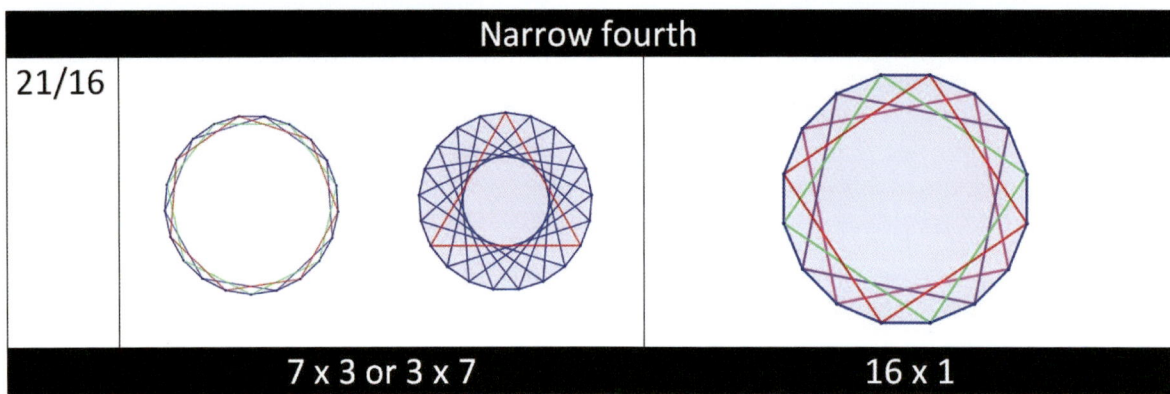
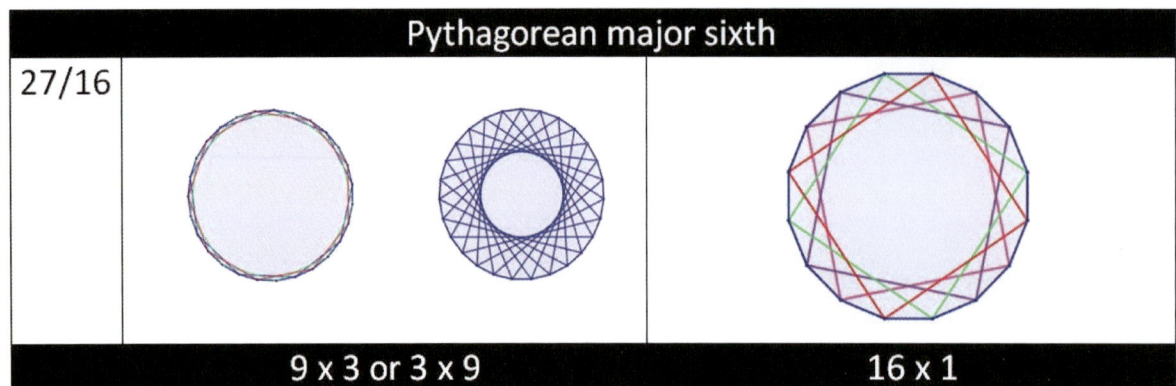

In these ratios the numbers on the right are all regular numbers, but the ones on the left are not regular in all of them. In musical intervals, regular numbers don't sound good with non-regular numbers.

11/8 sounds very odd. 11 represents a hendecagon.
13/8 sounds very odd. 13 represents a tridecagon.
21/16 sounds slightly odd. 21 contains 3 septagons or 7 triangles.
27/16 sounds quite **GOOD.** 27 contains 3 nonagons or 9 triangles.

11, 13 are not regular numbers and don't contain any smaller regular polygons, they sound very unpleasant.

21 contains 7 triangles and sounds slightly better than 11 and 13, but 7 is not a regular amount so it still sounds a bit strange. These 11, 13 and 27 based intervals are not 5-limit intervals.

27 is a regular number, so 27 does sound fairly good in ratios.
27 contains 3 nonagons or 9 triangles. Triangles and Nonagons have regular number amounts of sides, and they appear here in regular number amounts. The fact that 27 is such a large number with no lower whole number octaves means that it does sound a bit strange though. But it is not as bad as 11, 13 and 21. This is a 5-limit interval.

The shapes / numbers in the ratios of the odd sounding intervals contain non-regular number sided polygons, or non-regular amounts of regular sided polygons. This makes perfect sense because these are not 5-limit intervals, and cannot be made with products of powers of 2, 3, and 5.

IRRATIONAL ANGLES

If you look at the degrees of the angles in the corners of the regular polygons, you will see that polygons without regular number amounts of sides (7, 11, 13 and 14-sided polygons) have long, irrational (infinitely long) numbers, while the ones with regular number amounts of sides have whole numbers.

The reason why these shapes have irrational numbers is because regular polygons divide a 360 degree circle into even parts. 7, 11, 13 and 14 however are not divisors of 360 (numbers that can divide 360 evenly without irrational numbers).

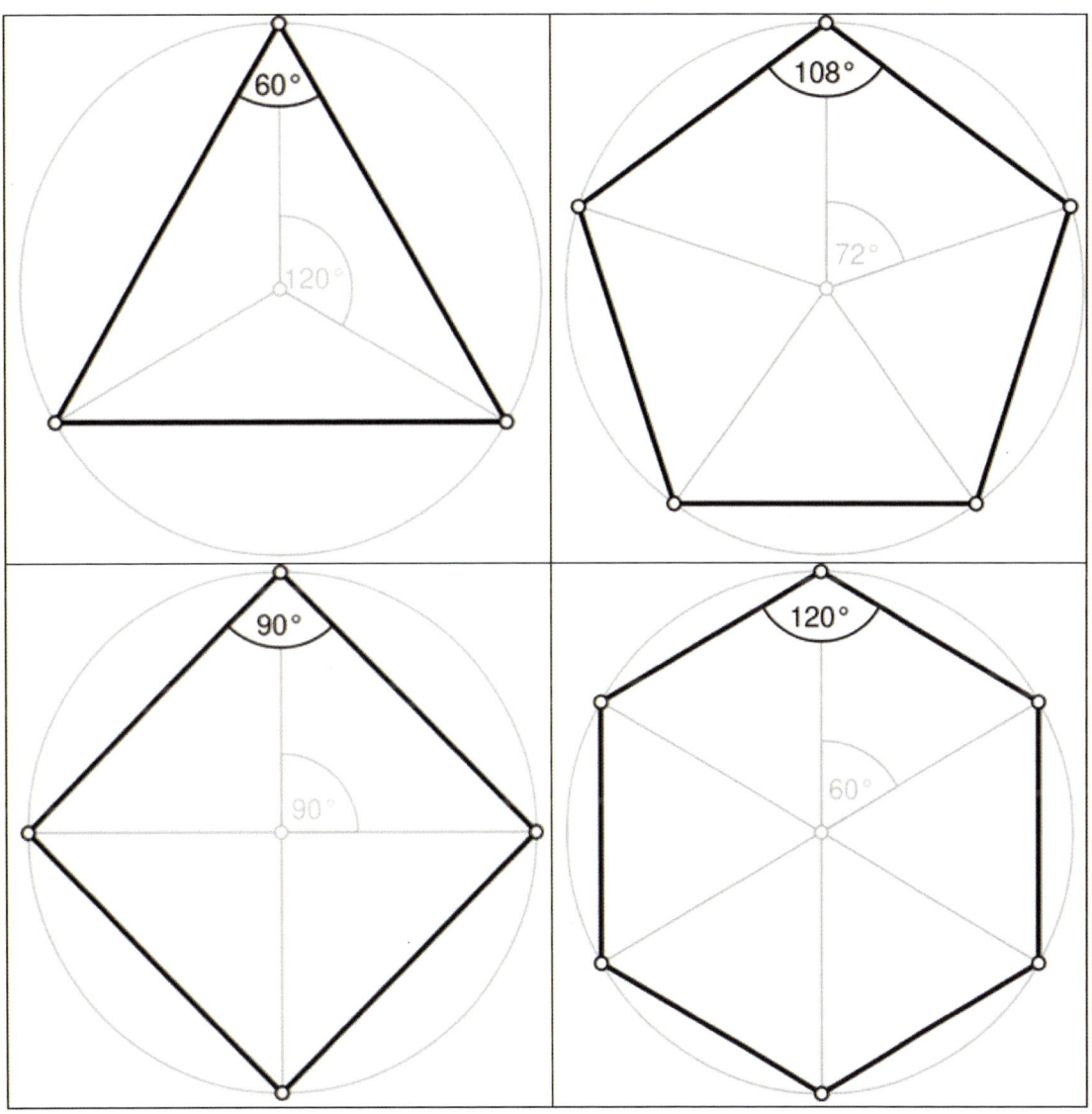

A regular triangle (3 sides) has 60 degrees in one corner.
A regular square (4 sides) has 90.
A regular pentagon (5 sides) has 108.
A regular hexagon (6 sides) has 120.
A regular septagon (7 sides) has 128.5714285714285714285...
A regular octagon (8 sides) has 135.
A regular nonagon (9 sides) has 140.
A regular decagon (10 sides) has 144.
A regular hendecagon (11 sides) has 147.2727272727272727...
A regular dodecagon (12 sides) has 150.
A regular tridecagon (13 sides) has 152.30769230769230769...
A regular tetradecagon (14 sides) has 154.285714285714285...
A regular pentadecagon (15 sides) has 156.
A regular hexadecagon (16 sides) has 157.5.

It is interesting how these irrational numbers are connected to the odd sounding intervals while the rational numbers are connected to good sounding ones.

You can extend the geometry of the regular polygons with odd numbers of sides by joining the corners. This divides the internal angle into smaller ones, making polygrams inside the polygons. With a pentagon you will get a pentagram. With the heptagon, a heptagram etc.

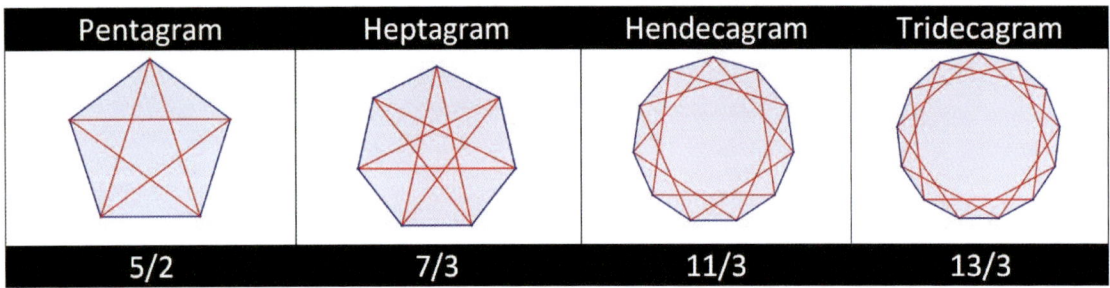

In some polygons you can make more than one type polygram. These are written as ratios. For example; the pentagram in the pentagon above is written as 5/2 because every second corner in a 5-sided polygon was joined. The heptagram is written as 7/3 because every third corner in a 7 sided polygon was joined. You can also make a 7/2 heptagram. All polygram ratios work in the same way.

You only get a 5/2 pentagram, and its internal angles are the regular numbers 36, 72, and 108.

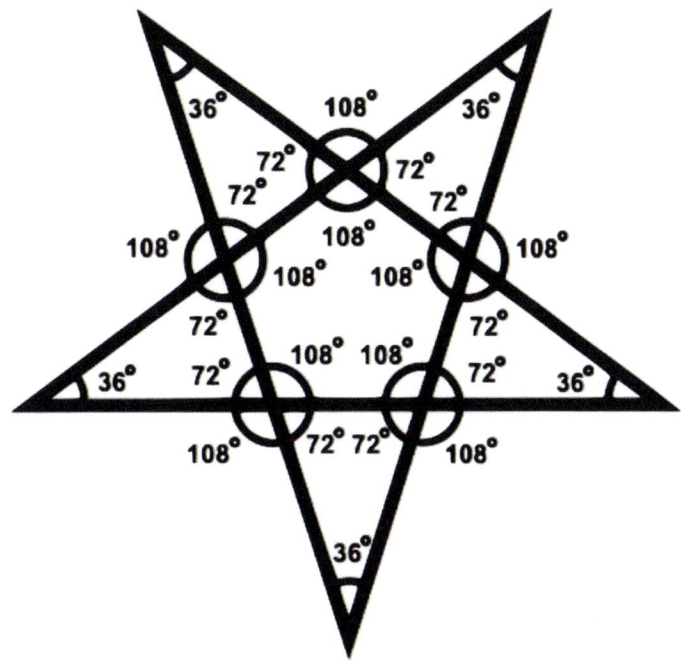

This is an octave (36 to 72) and a perfect fifth (72 to 108); the notes D, D and A in Ptolemy's intense diatonic scale with 24 Hz as reference pitch.

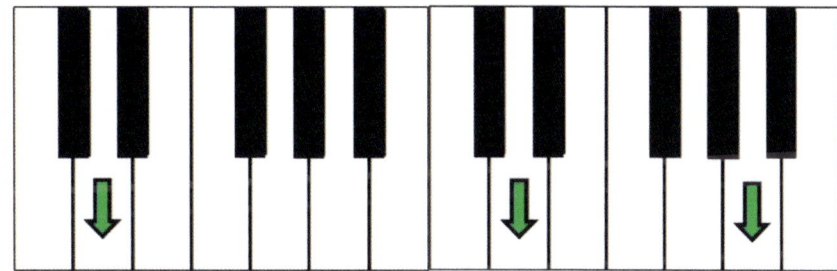

All polygons with regular number amounts of sides that I have seen have rational numbers in their polygrams. Even the nine-sided nonagram has 20 or 100 degrees in its main angle, depending on what kind of nonagram it is. This must be so because you are dividing that first angle into smaller regular parts, which seems to preserve the number's mathematical properties.

If you make polygrams inside 7, 11, 13 or 14-sided polygons, though, you just get more irrational numbers as the properties of the number in the first angle seems to fractal down through its divisions. So, while the regular number sided pentagram and the nonagrams do look similar to the non-regular number sided 7, 11, and 13 sided polygrams, their mathematical and musical properties are in fact very different.

Dividing the Sphere

The polygons and polygrams in the last chapter were equally dividing a circle. Divisions of 7, 11, 13 and 14 made irrational numbers mathematically, while the same numbers made odd sounding musical intervals. Other than that they were not physically different from ones made with regular numbers than have the good sounding intervals. If we move up a dimension and equally divide a sphere, however, things get a lot more interesting. Then the mathematical and musical properties of numbers become physical.

You only actually get 5 different ways to divide a sphere exactly evenly. The 5 three dimensional shapes you get when you do this have been known and studied for thousands of years for their extreme beauty and simplicity. Today we call them the Platonic solids.

Platonic solids (Dice of the Gods)

A Platonic solid is a regular, convex polyhedron. They are made from identical-in-shape-and-size, regular polygonal faces with the same number of faces meeting at each vertex. All corners must also touch the outside of a sphere, dividing it into even parts. Basically, every side and angle in each shape is exactly the same, making them perfectly uniform. Because of this, these are also the only fair rolling dice that can be made.

Plato believed that everything was made from tiny platonic solids. This has not been proven to be true, although nobody really knows exactly what is going on at the tiniest levels. Some people say that tiny atomic and subatomic particles might actually move in a way that traces out these shapes.

Plato connected the 5 Platonic solids to the 5 elements:

Hexahedron (cube) = Earth
Icosahedron = Water
Tetrahedron = Fire
Octahedron = Air
Dodecahedron = Aether

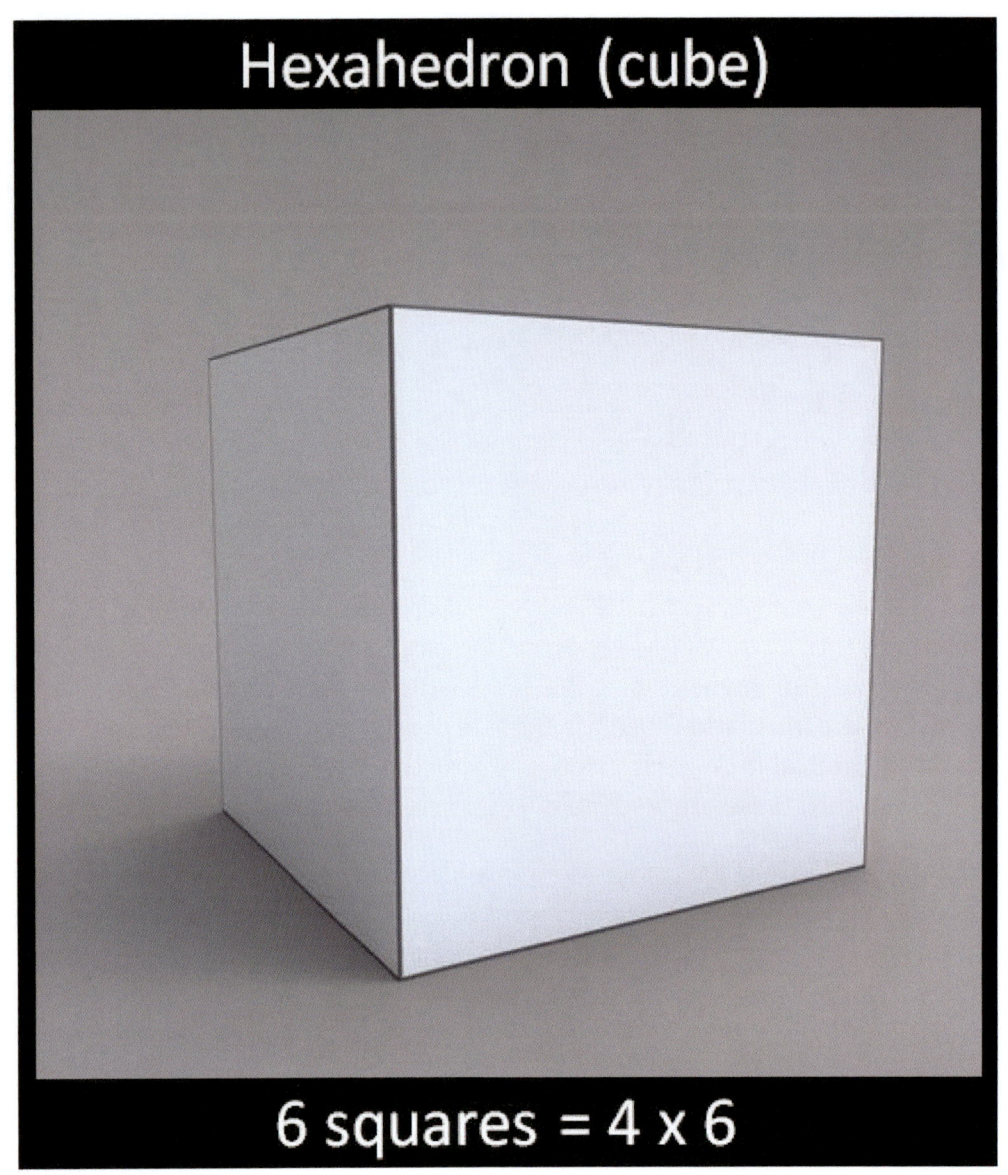

Icosahedron

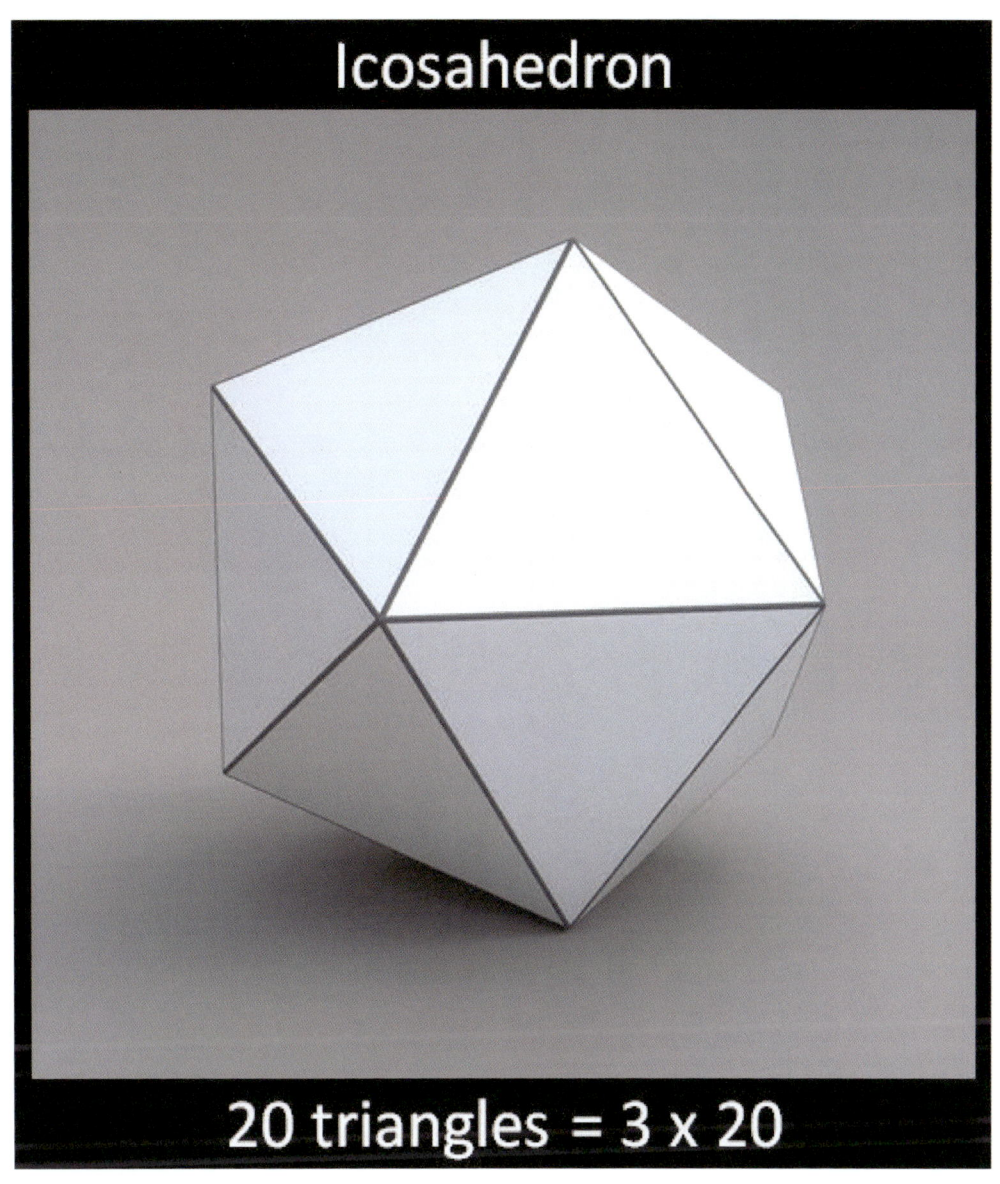

20 triangles = 3 x 20

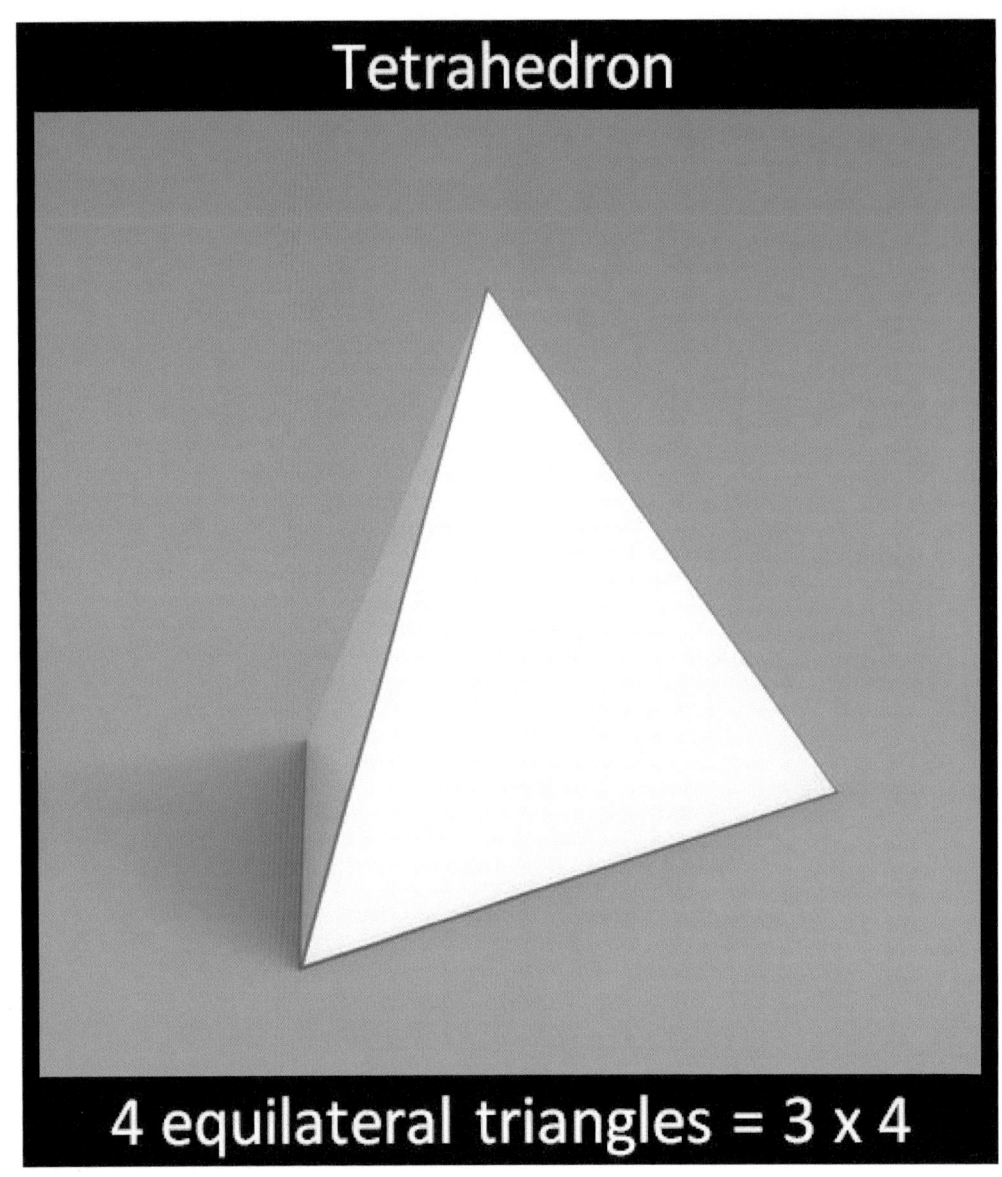

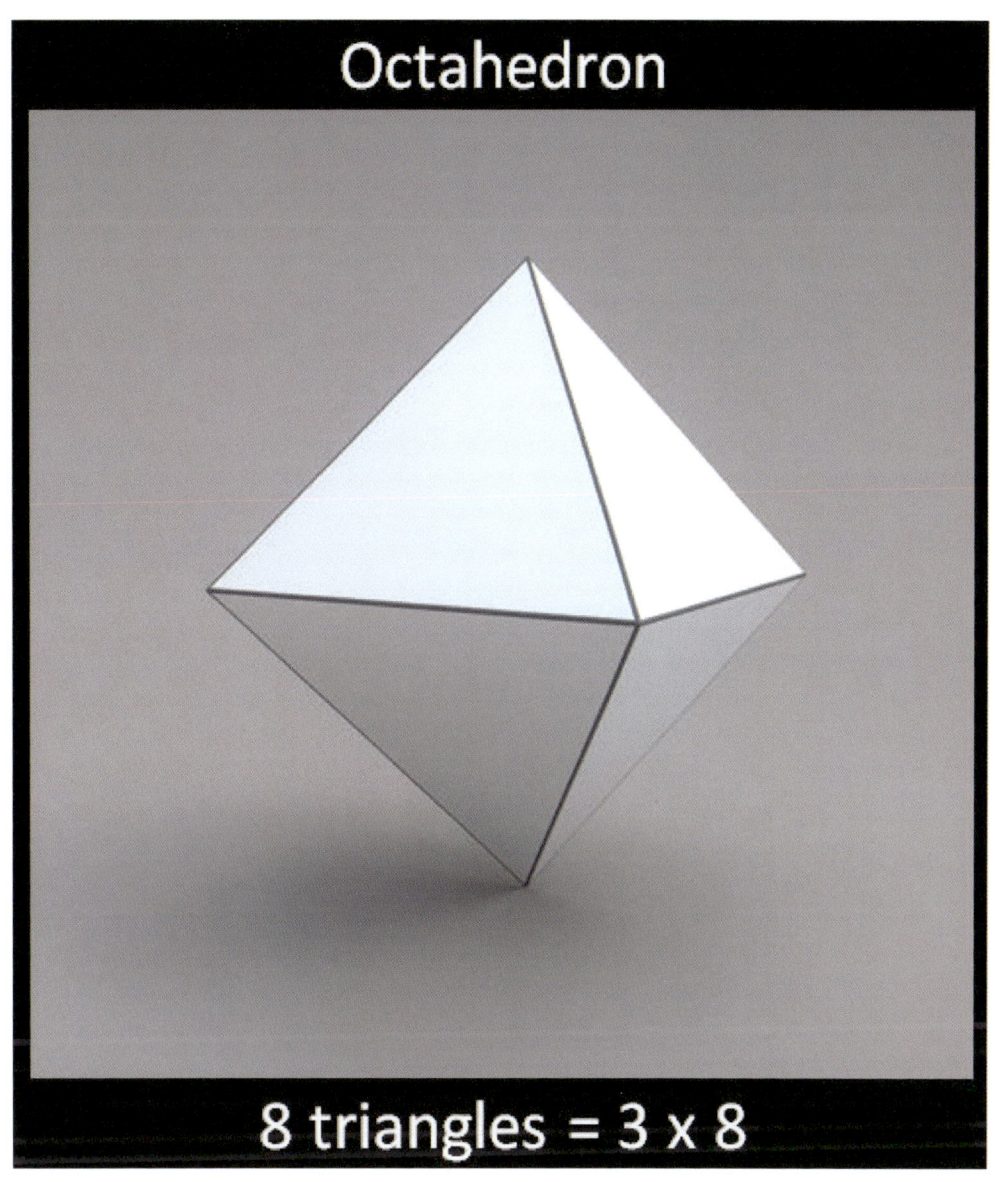

Octahedron

8 triangles = 3 x 8

Dodecahedron

12 Pentagons = 5 x 12

As you can see they are made from only triangles, squares, or pentagons. You cannot use 7, 11, 13 or 14-sided polygons to evenly divide a sphere. In fact, no other polygons can evenly divide a sphere. Only triangles, squares or pentagons can do it.

You can't use any random amount of triangles, squares, or pentagons to divide a sphere either. You can only use regular number amounts of them. So, using 7, 11 or 13 of them will not work. This is very musical because a regular number multiplied by another regular number is how you can make 5-limit music intervals. In the images above you can see how the amount of sides in each polygon is always a regular number, as is the amount of them in the solid.

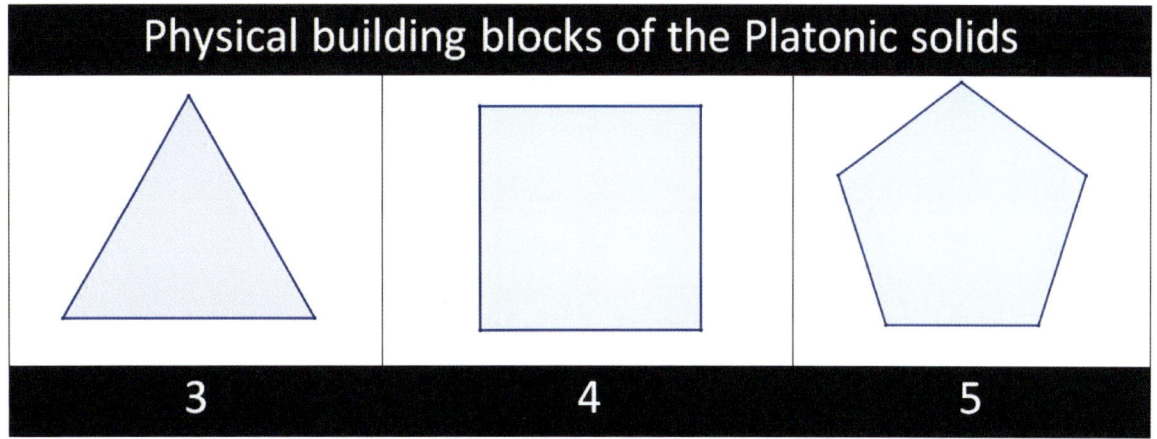

In the levitated drop of water, triangles, squares, and pentagons represent the 3-4-5 pure major chord. So, you have a very nice resonance between the building blocks of the platonic solids.

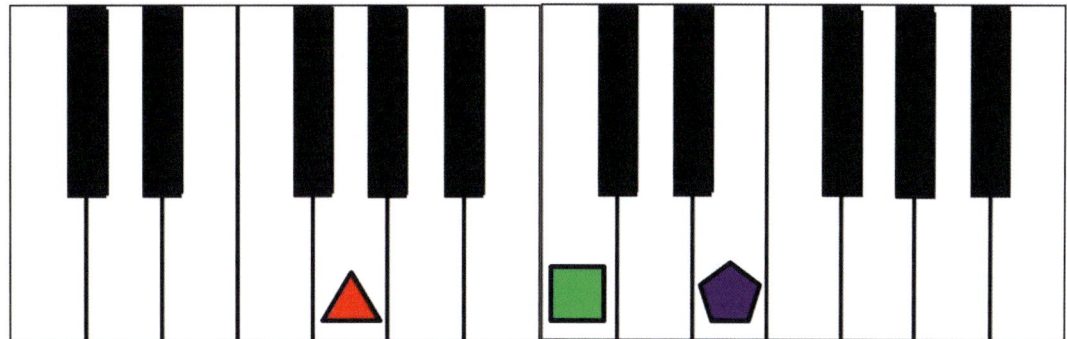

3, 4, and 5 have their own amazing geometry, revealing the Pythagorean theorem. 3 squared + 4 squared = 5 squared. You will find that throughout sacred geometry, nice sounding musical intervals always mirror the simplest and most useful shapes.

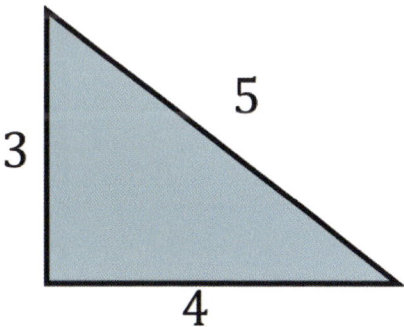

Remember that a 5 limit music scale is made from products of powers of 2, 3 and 5, and since 2 x 2 = 4, the numbers 3, 4 and 5 from which the platonic solids are made are also made from products of powers of 2, 3 and 5. Anything that can be done with 3, 4 and 5 can also be done with 2, 3 and 5, so things made from 3, 4 and 5 (triangles, squares and pentagons) still fall into the 5 limit rule mathematically.

This makes me wonder... sound really does move away from its source, in all directions, in a spherical shape. So, maybe the fact that regular number intervals sound good together, while non-regular number intervals don't sound good with them, is related to the way they fit or don't together as spherical sound waves in the air?

I have searched the web for info on vibrating spheres and platonic solids but only found two paragraphs. I found the exact same two paragraphs, word for word, on a million websites... but no other info on this idea. Buckminster Fuller was a well-respected person who invented amazing things like the widely-known geodesic dome. Carbon molecules known as fullerenes were later named by scientists for their structural and mathematical resemblance to geodesic spheres. So, I do think this may be true.

"The late Buckminster Fuller (1895- 1983 was the first to discover that a relation between musical frequencies and geometrical forms exists. He used a balloon submerged in blue dye and vibrated it with frequencies from the diatonic musical scale (the 7 white keys from the piano): as a result of wave interference, marvellous two dimensional arrangements appeared on its surface.

In the experiment conducted by Fuller's students, a spherical balloon was dipped in dye and pulsed with "pure" sound frequencies, known as the "Diatonic" sound ratios.

A small number of evenly-distanced nodes would form across the surface of the sphere, as well as thin lines that connected them to each other. If you have four evenly spaced nodes, you will see a tetrahedron. Six evenly spaced nodes form an octahedron. Eight evenly spaced nodes form a cube. Twenty evenly spaced nodes form the dodecahedron, and twelve evenly spaced nodes form the icosahedron".

"Dr. Hans Jenny (his student)conducted a similar experiment, wherein a droplet of water contained a very fine suspension of light-colored particles, known as a colloidal suspension. When this roughly spherical droplet of particle-filled water was vibrated at various Diatonic musical frequencies, the Platonic Solids would appear inside, surrounded by elliptical curving lines that would connect their nodes together".

This all makes sense, because in cymatics the plate gets divided into equal portions by vibration nodes. So, with spherical cymatics the sphere must also be divided evenly by nodes, and only the platonic solids can do that, because all of their sides are equal in length.

The fact that they used "pure diatonic frequencies" is interesting, as that is exactly what I would have used if I was them. The fact that they speak of *"wave interference"* also makes me think they were playing intervals, and not just one note at a time.

ARCHIMEDEAN SOLIDS

If you use more than one type of regular polygon together in one shape, you can make the Archimedean solids. Obviously these are named after Archimedes and only 13 of them exist. The Archimedean solids are referred to as the semiregular polyhedra, because they are not 100% regular like the platonic solids are. Although they are not 100% regular, all of the edges in these are also exactly the same length.

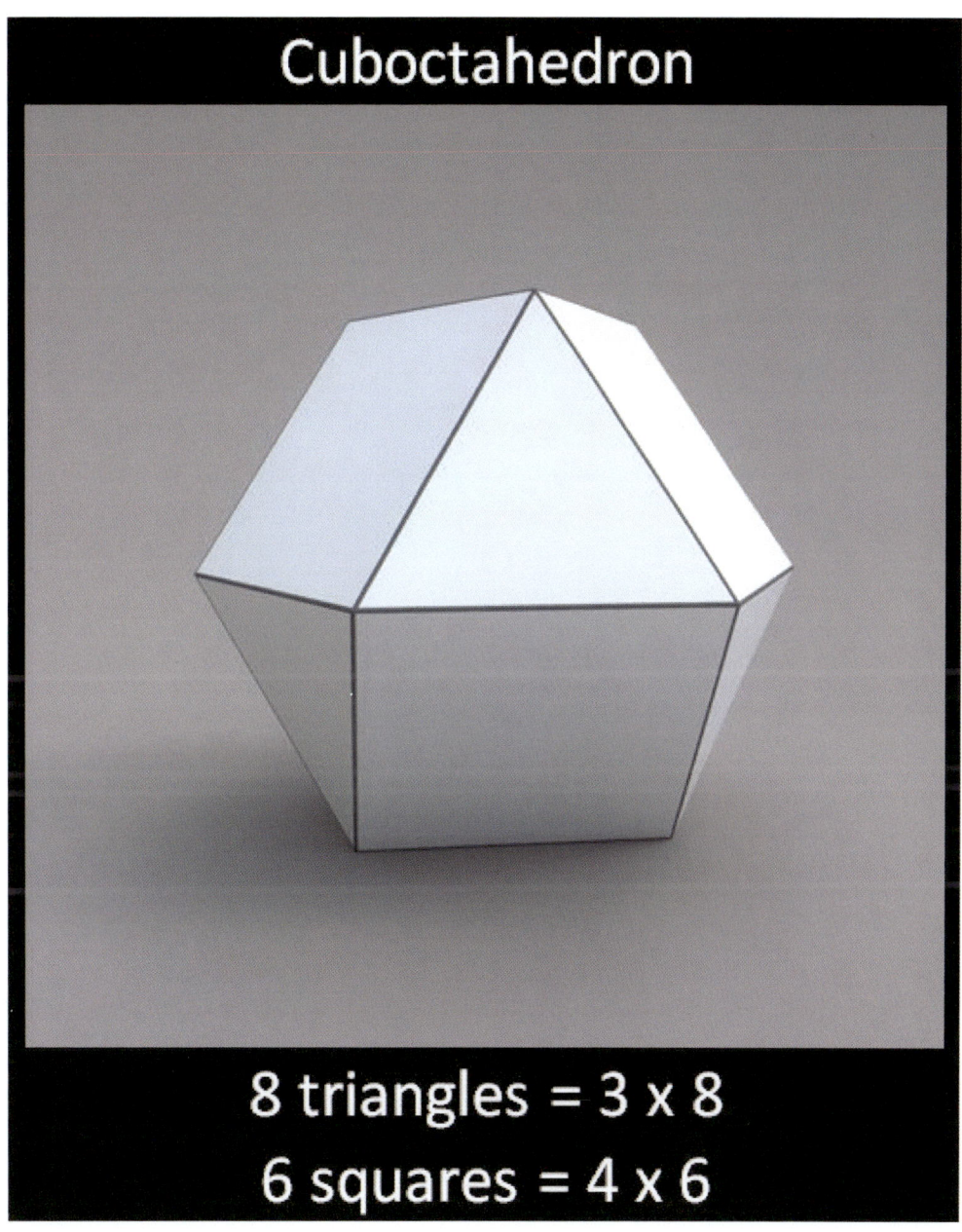

Cuboctahedron

8 triangles = 3 x 8
6 squares = 4 x 6

Icosidodecahedron

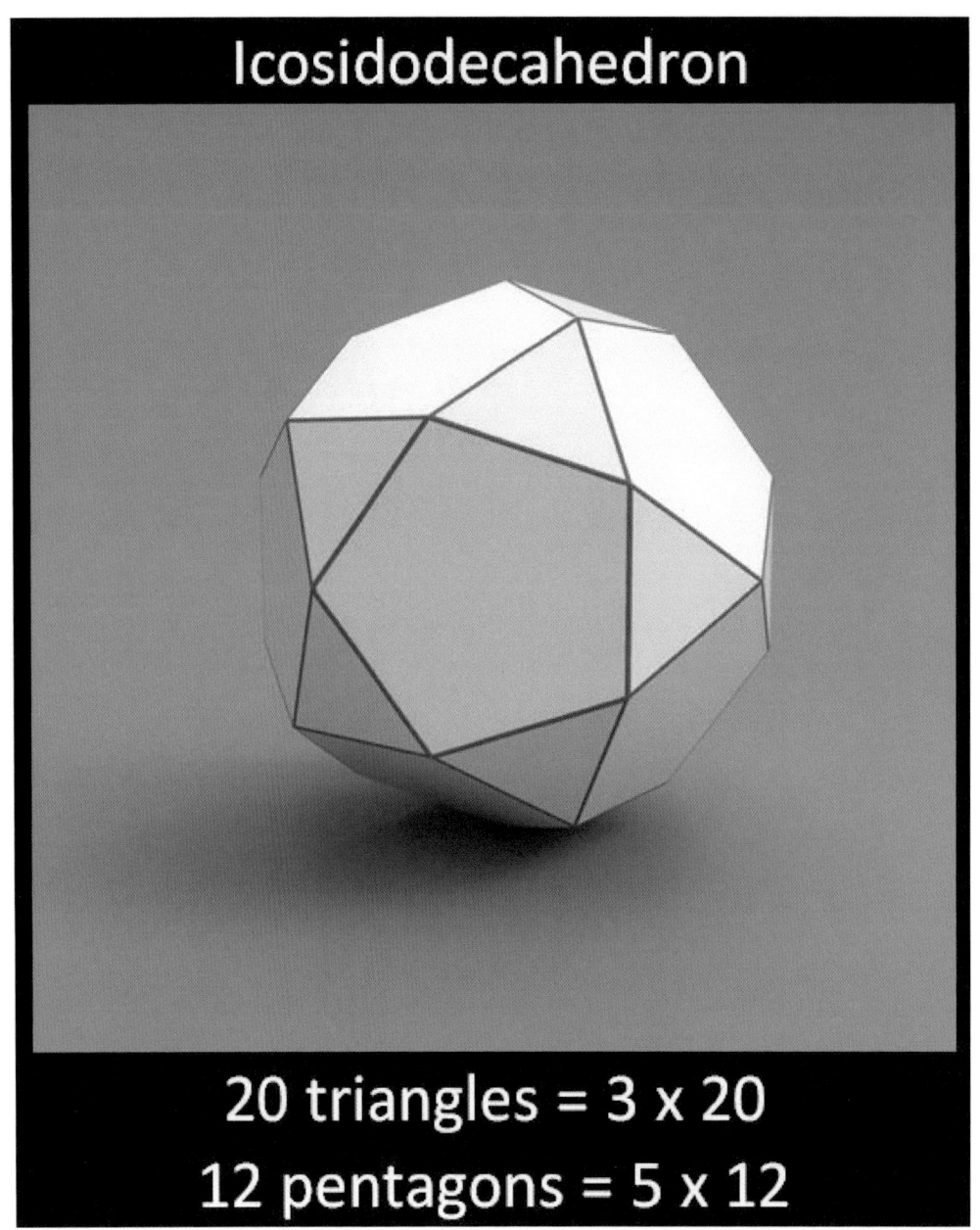

20 triangles = 3 x 20
12 pentagons = 5 x 12

Rhombicosidodecahedron

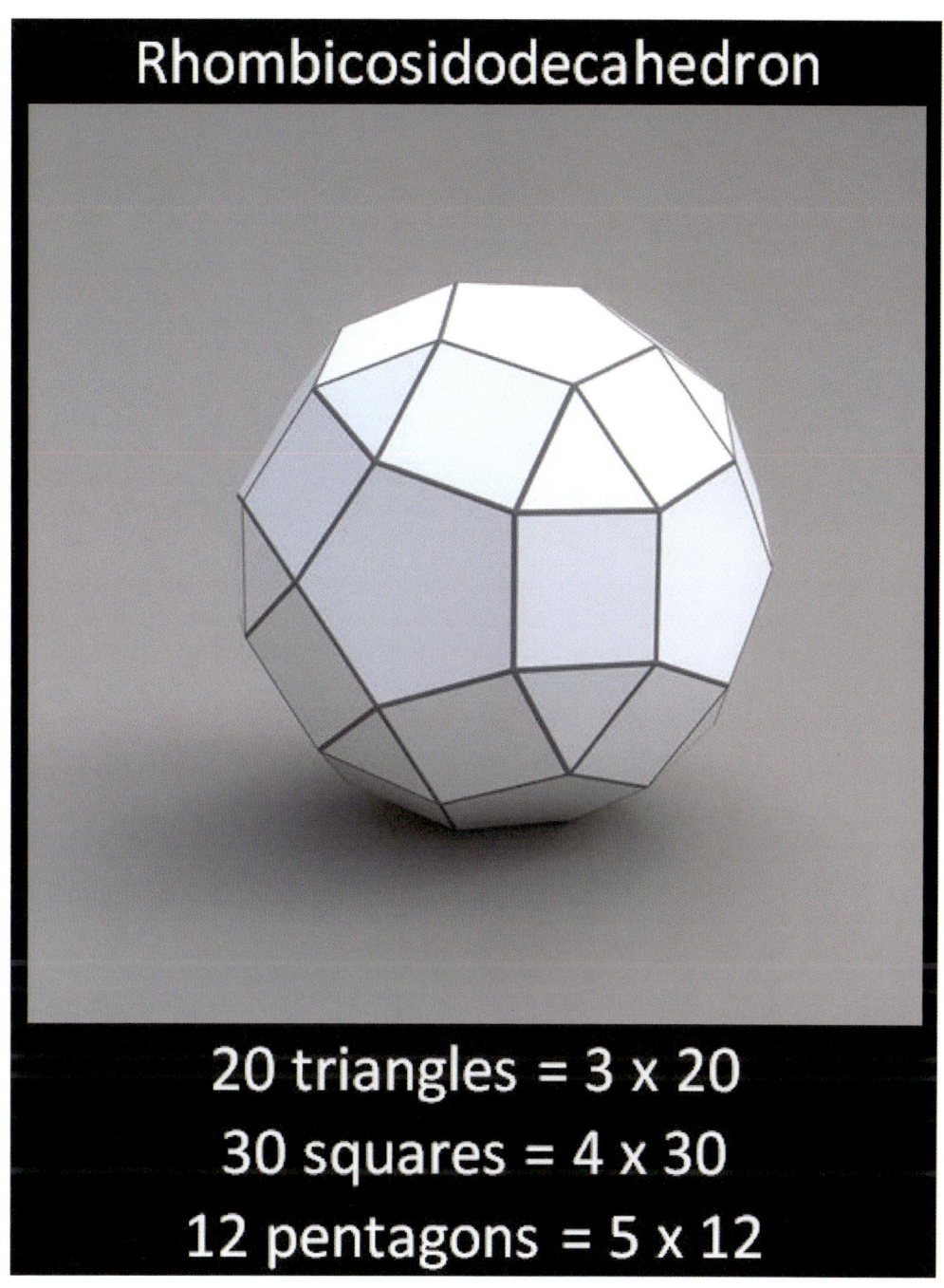

20 triangles = 3 x 20
30 squares = 4 x 30
12 pentagons = 5 x 12

Rhombicuboctahedron

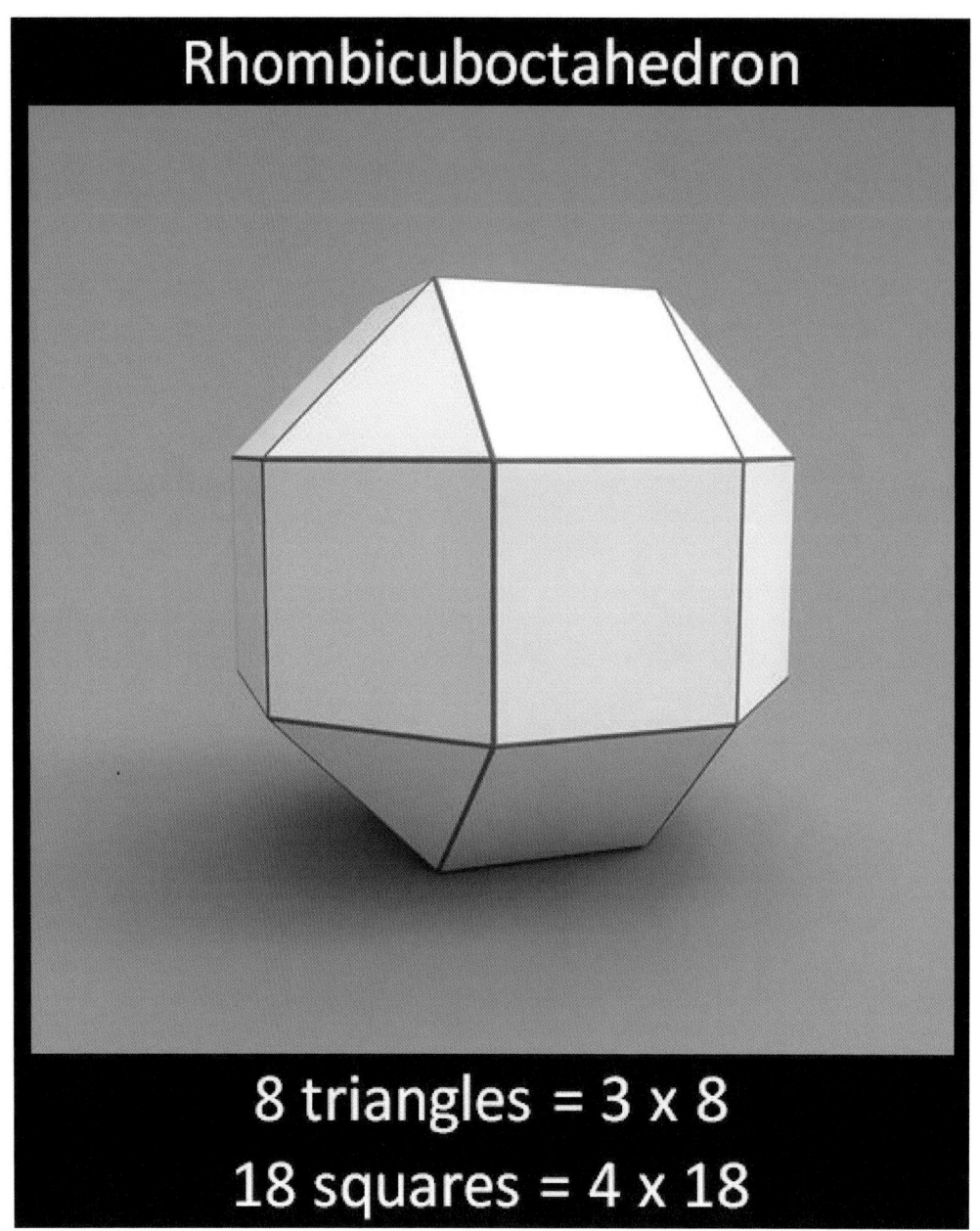

8 triangles = 3 x 8
18 squares = 4 x 18

Snub cube

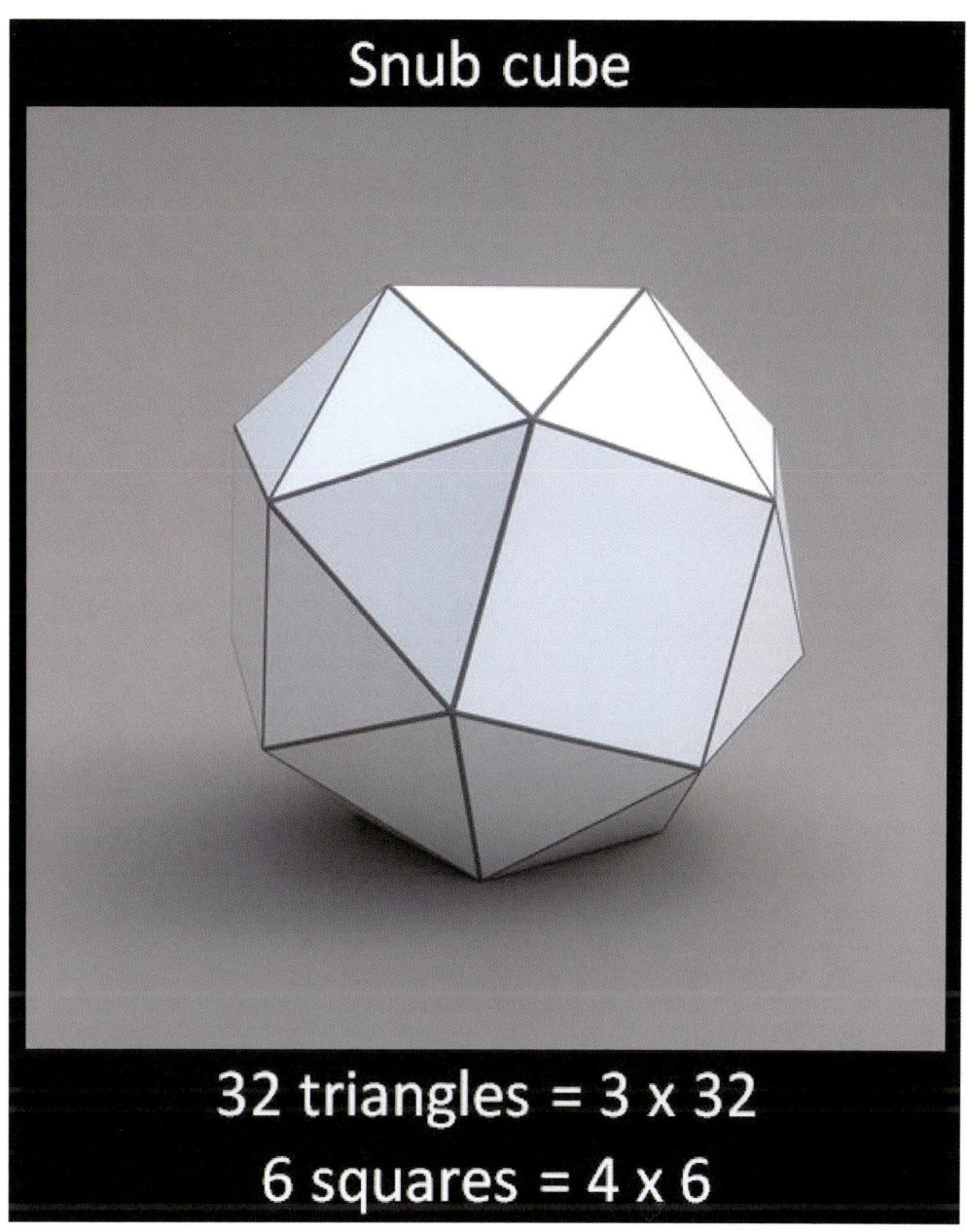

32 triangles = 3 x 32
6 squares = 4 x 6

Snub dodecahedron

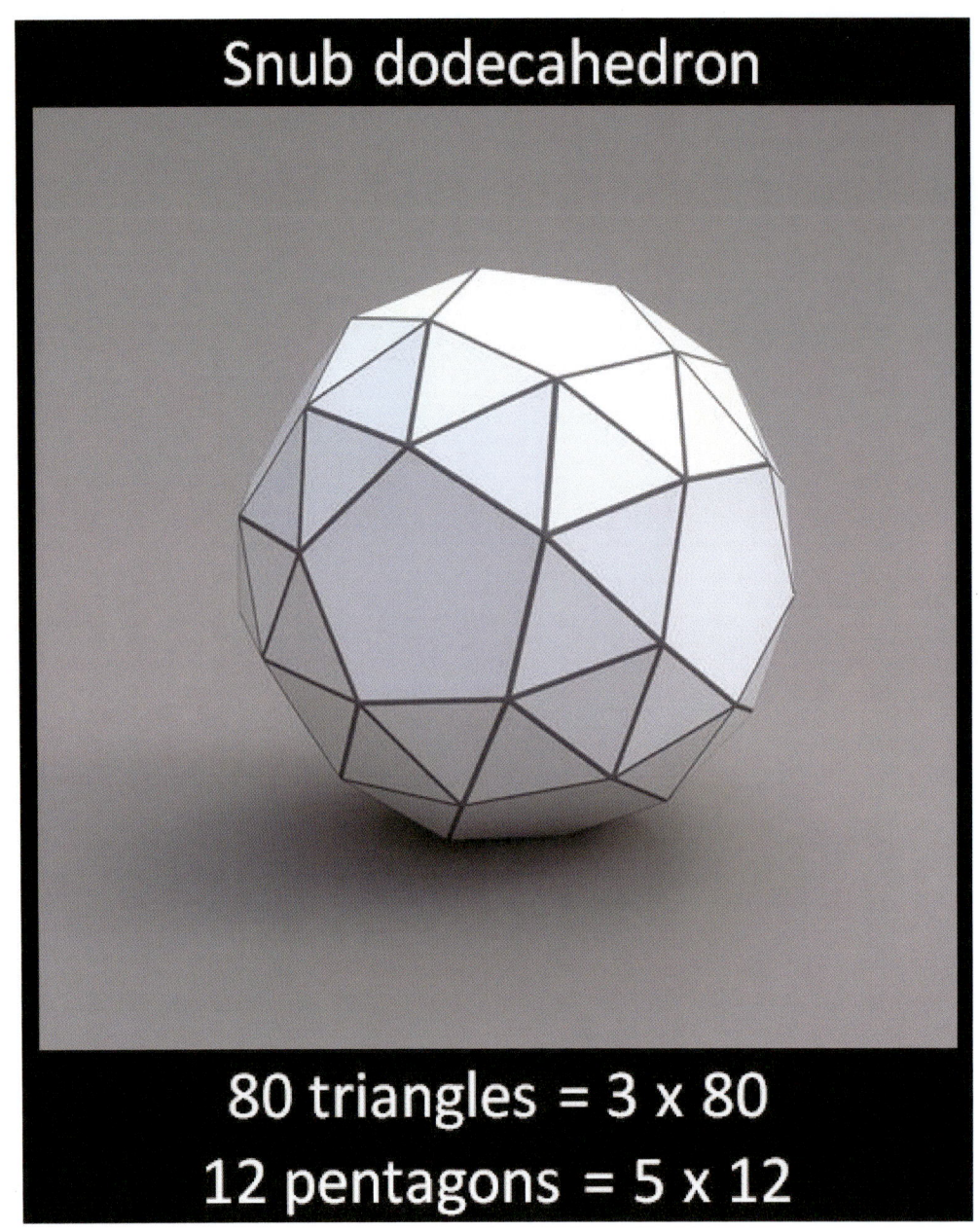

80 triangles = 3 x 80
12 pentagons = 5 x 12

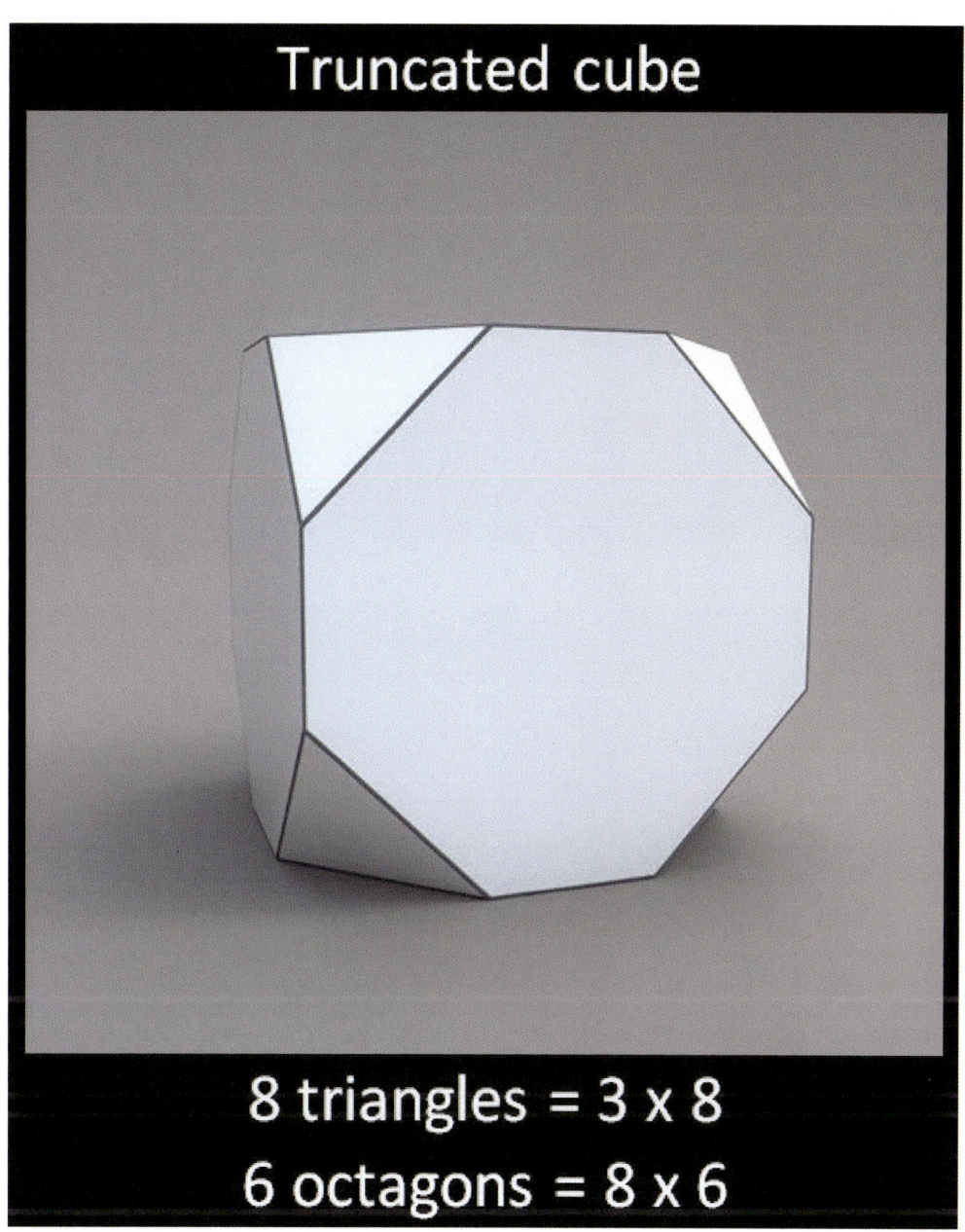

Truncated cube

8 triangles = 3 x 8
6 octagons = 8 x 6

Truncated cuboctahedron

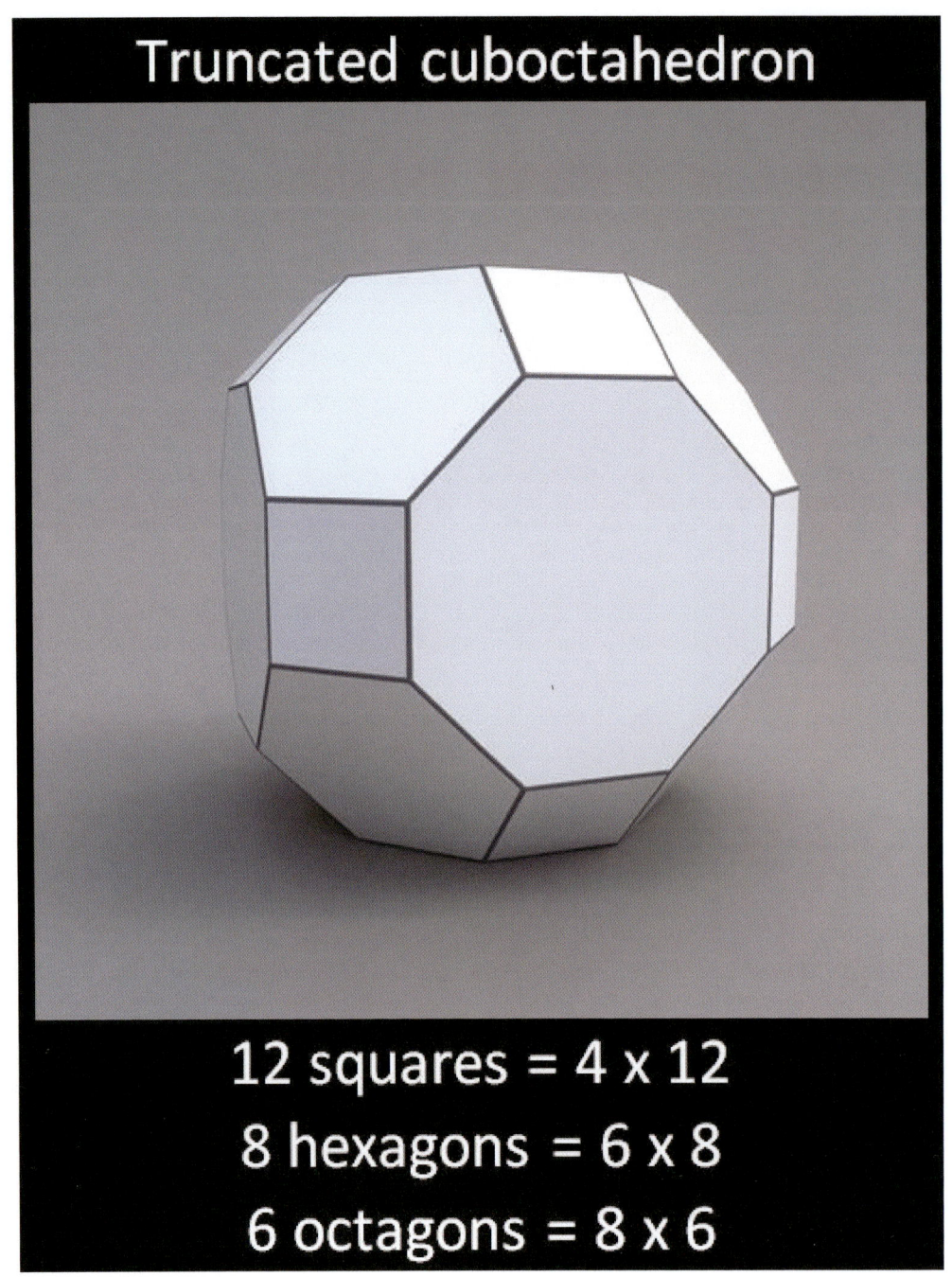

12 squares = 4 x 12
8 hexagons = 6 x 8
6 octagons = 8 x 6

Truncated dodecahedron

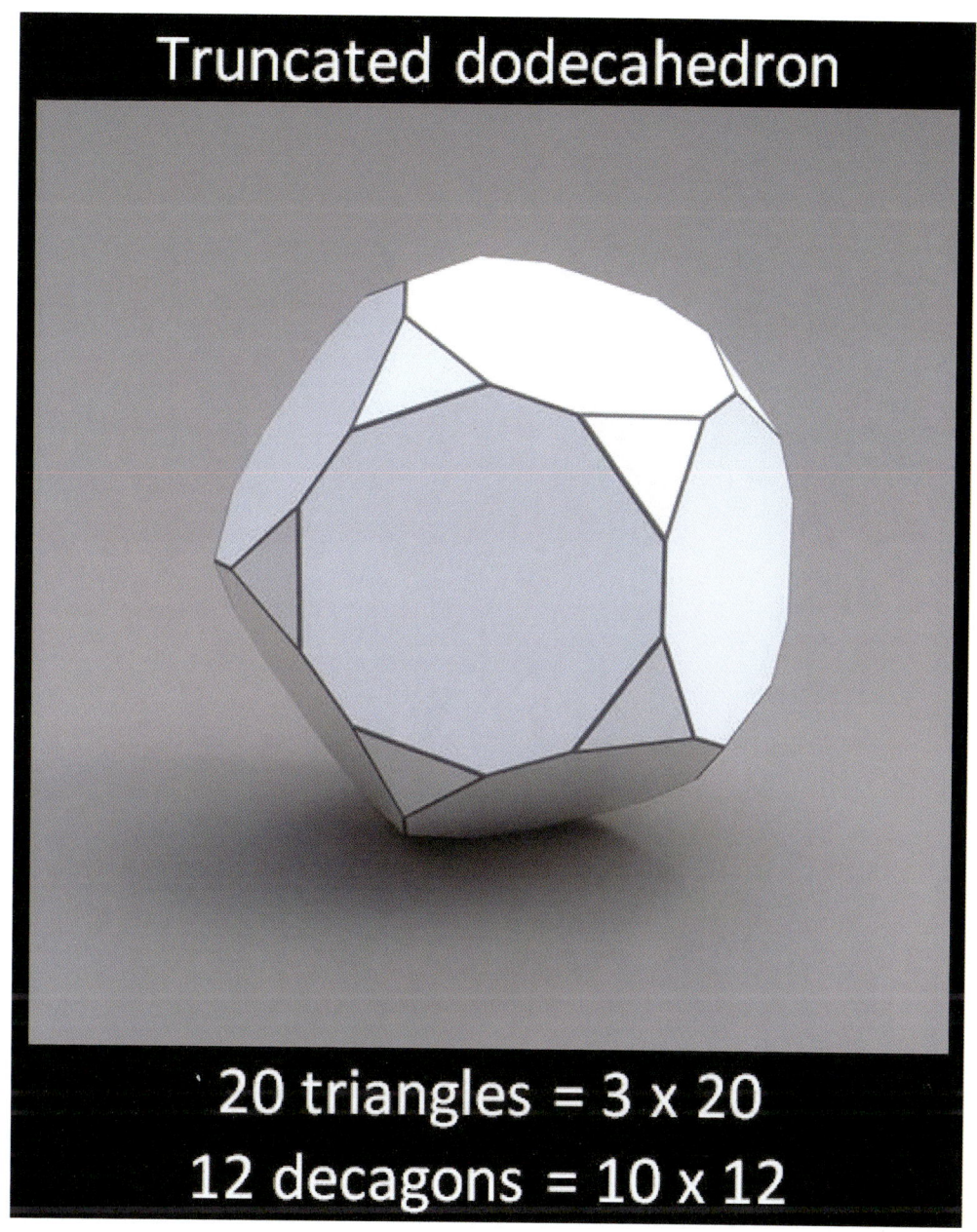

20 triangles = 3 x 20
12 decagons = 10 x 12

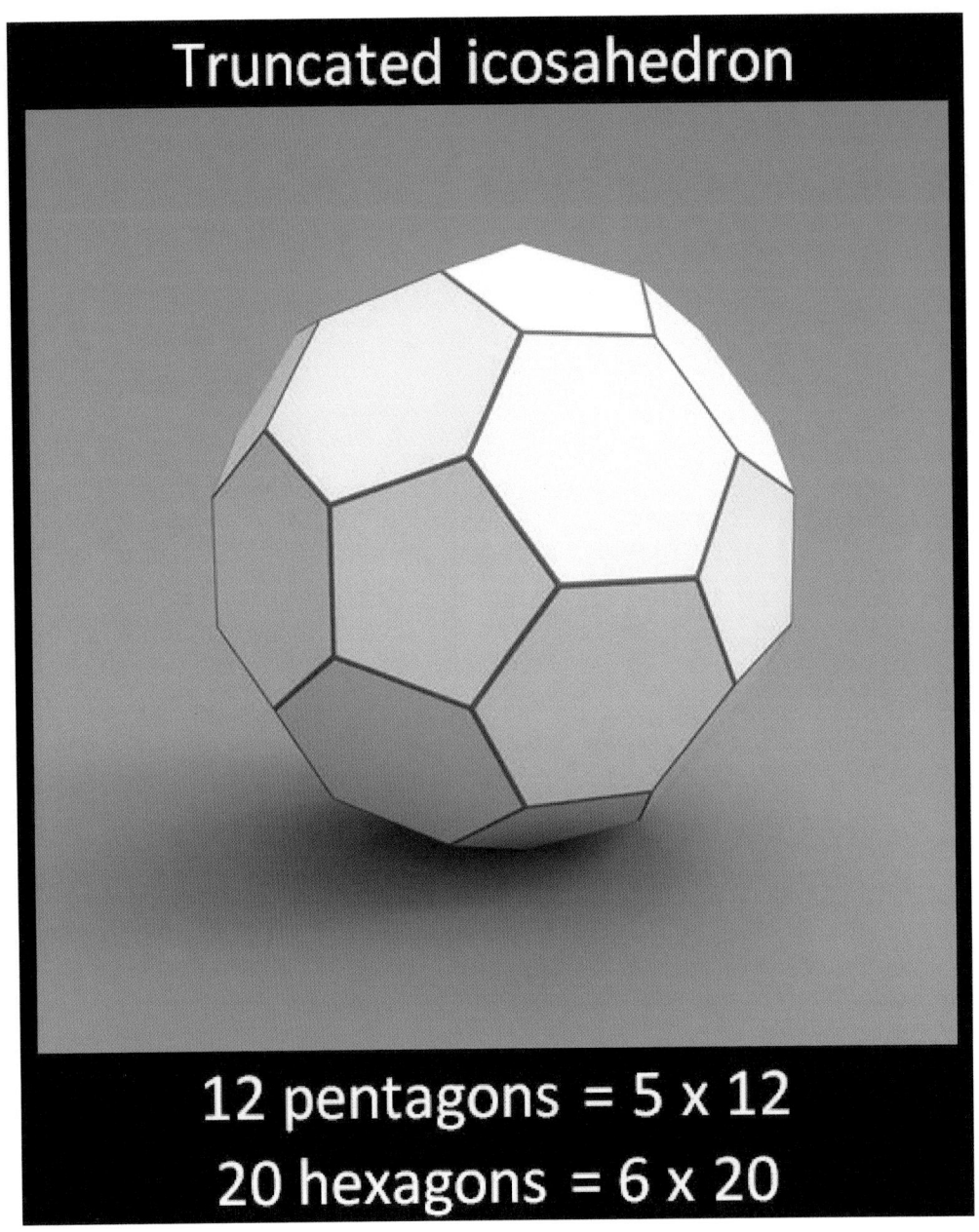

Truncated icosidodecahedron

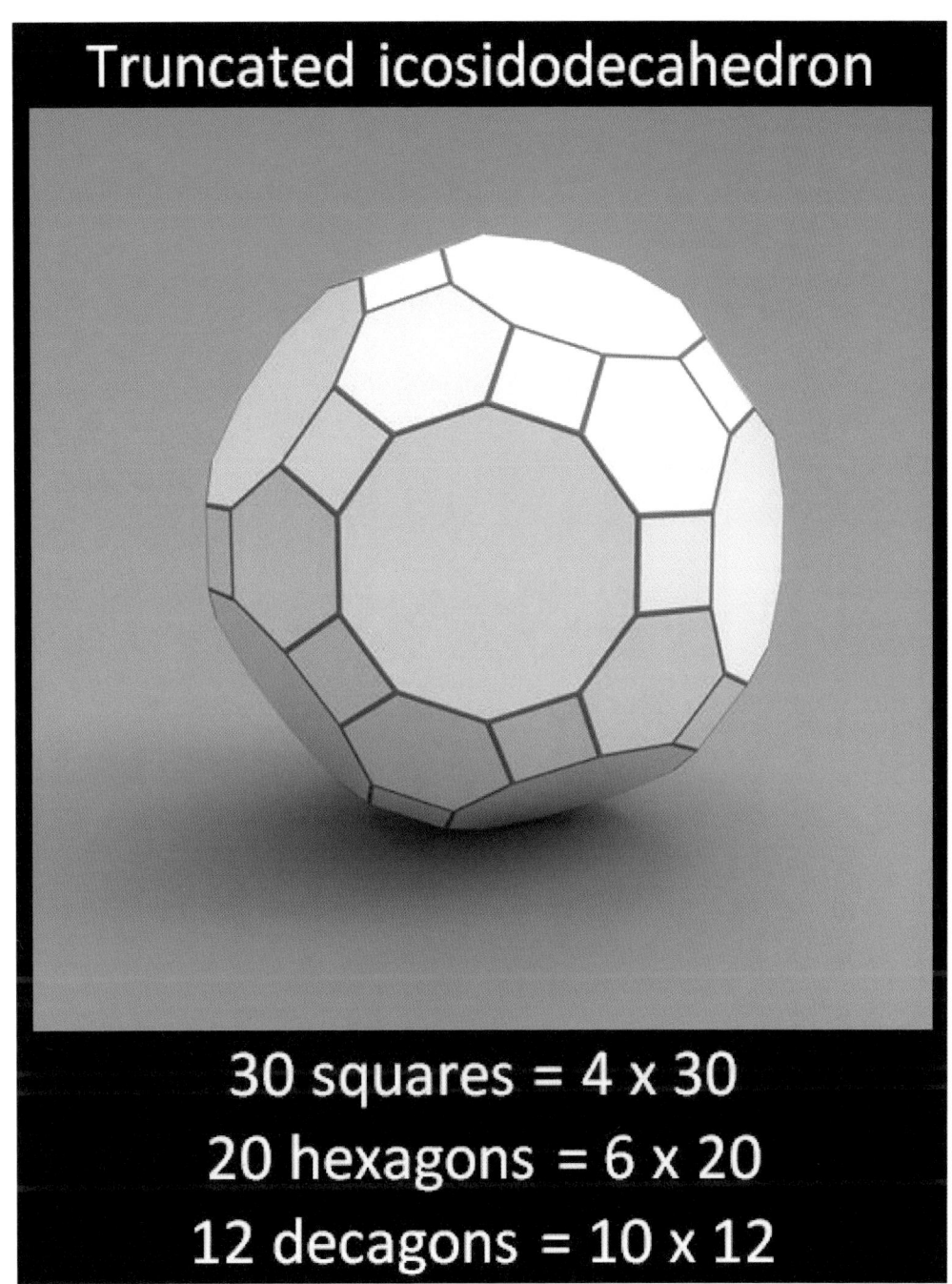

30 squares = 4 x 30
20 hexagons = 6 x 20
12 decagons = 10 x 12

Truncated octahedron

6 squares = 4 x 6
8 hexagons = 6 x 8

Truncated tetrahedron

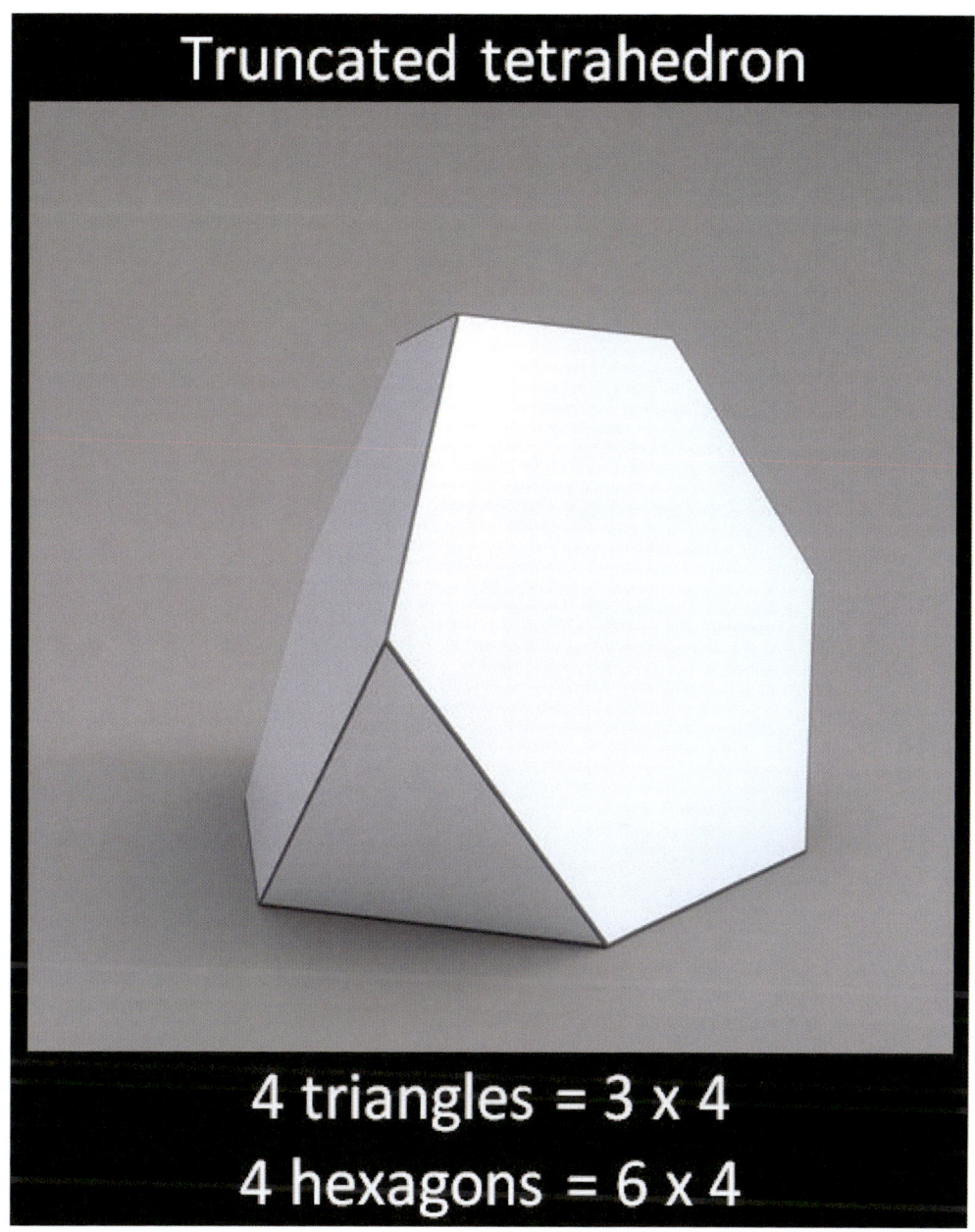

4 triangles = 3 x 4
4 hexagons = 6 x 4

As you can see, Archimedean solids contain regular number amounts of only triangles, squares, pentagons, hexagons, octagons and decagons (3, 4, 5, 6, 8 and 10 sided polygons), which themselves have regular number amounts of sides. You also cannot use 7, 11, or 13-sided polygons or other non-regular number amounts of polygons to make these semi-regular polyhedrons. Since the definition of a 5-limit interval is that the ratios must contain only regular numbers, the Archimedean solids are essentially also "5-limit polyhedrons".

A hexagon is an octave of a triangle, which is why it contains 2 triangles. In the same way an octagon is an octave of a square, and a decagon is an octave of a pentagon. So mathematically the Archimedean solids have the same basic building blocks as the Platonic solids, the triangle, the square and the pentagon (3, 4 and 5). Since 3 and 5 are already prime numbers, while the prime factors of 4 are 2 x 2, these also have the prime factors 2, 3 and 5 mathematically.

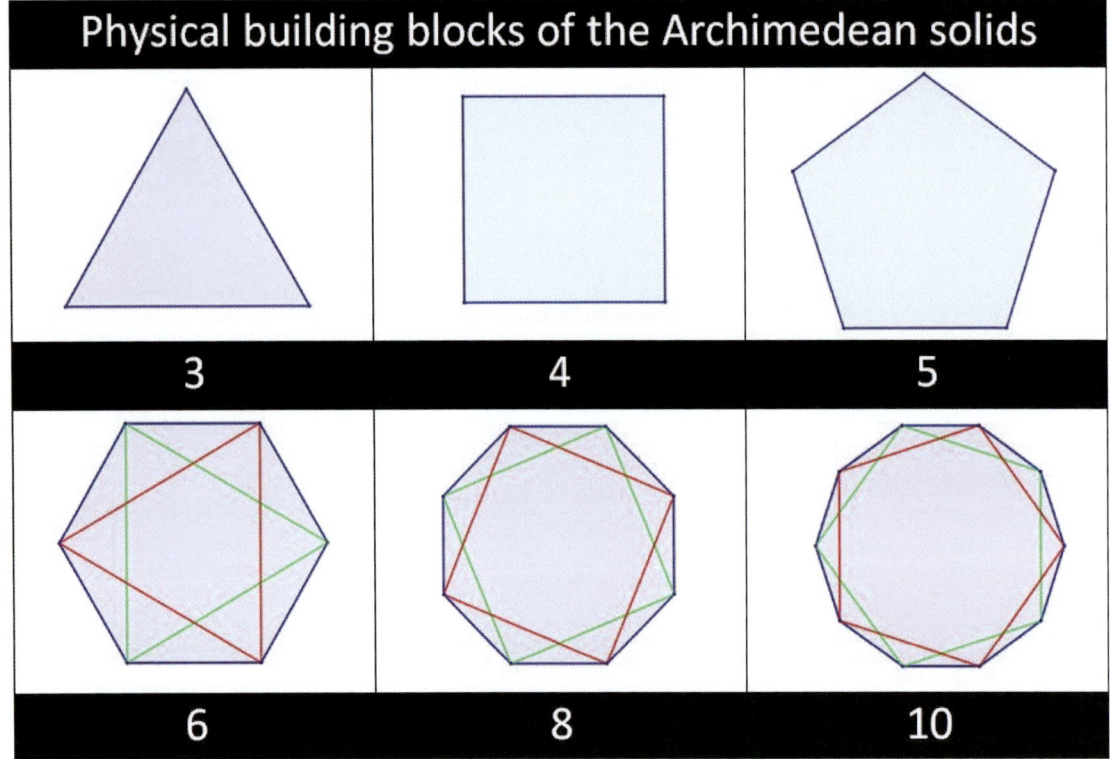

Musically triangles, squares, pentagons, hexagons, octagons and decagons represent the pure 3-4-5 major chord over 2 octaves, so it is 3-4-5-6-8-10. You can try it on a piano to hear for yourself how musical the intervals found in the angles of the Archimedean solids are.

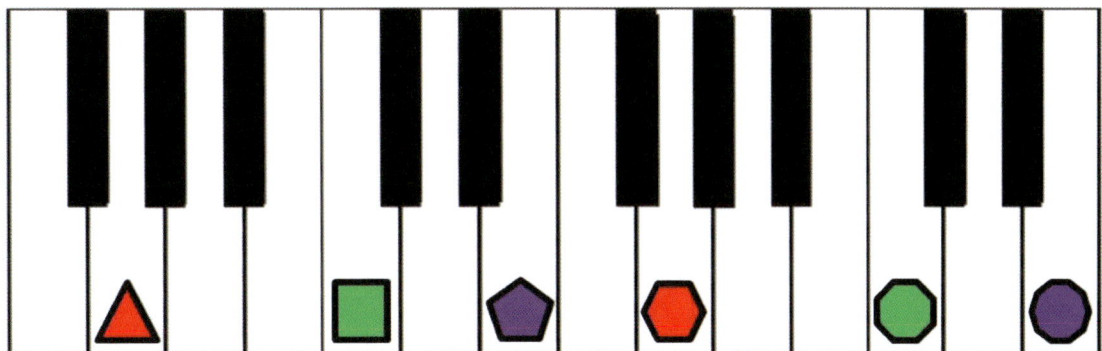

You can find 7 intervals between these notes: the octave (C to C), major sixth (G to E), minor sixth (E to C), perfect fifth (C to G), perfect fourth (G to C), major third (C to E) and minor third (E to G). You can use the following chart with the piano image above to hear which of these intervals are between the polygons in each Archimedean solid:

Archimedean solids				
Cuboctahedron	triangles	squares		3 - 4
Icosidodecahedron	triangles	pentagons		3 - 5
Rhombicosidodecahedron	triangles	squares	pentagons	3 - 4 - 5
Rhombicuboctahedron	triangles	squares		3 - 4
Snub cube	triangles	squares		3 - 4
Snub dodecahedron	triangles	pentagons		3 - 5
Truncated cube	triangles	octagons		3 - 8
Truncated cuboctahedron	squares	hexagons	octagons	4 - 6 - 8
Truncated dodecahedron	triangles	decagons		3 - 10
Truncated icosahedron	pentagons	hexagons		5 - 6
Truncated icosidodecahedron	squares	hexagons	decagons	4 - 6 - 10
Truncated octahedron	squares	hexagons		4 - 6
Truncated tetrahedron	triangles	hexagons		3 - 6

So, the Platonic and Archimedean solids could be called a type of 5-limit geometry. This limit is not imposed however; it is the physical limit of 7, 11, and 13-sided polygons, or those amounts of regular sided polygons not working in spherical 3D space at all. It is strange that the 5-prime-limit used in music to make nice sounding scales, and in maths to avoid irrational numbers, also represents a physical limit in 3D space.

As physical objects, most of the Platonic and Archimedean solids are very strong. One of the reasons why they are so strong is because they all have edges of the exact same length. This minimizes the places where weak spots can exist.

The cube is widely used in construction; the stone bricks used to make pyramids and other buildings are usually cubes or rectangles.

Cubes are also often found in crystal formations, for example this salt crystal:

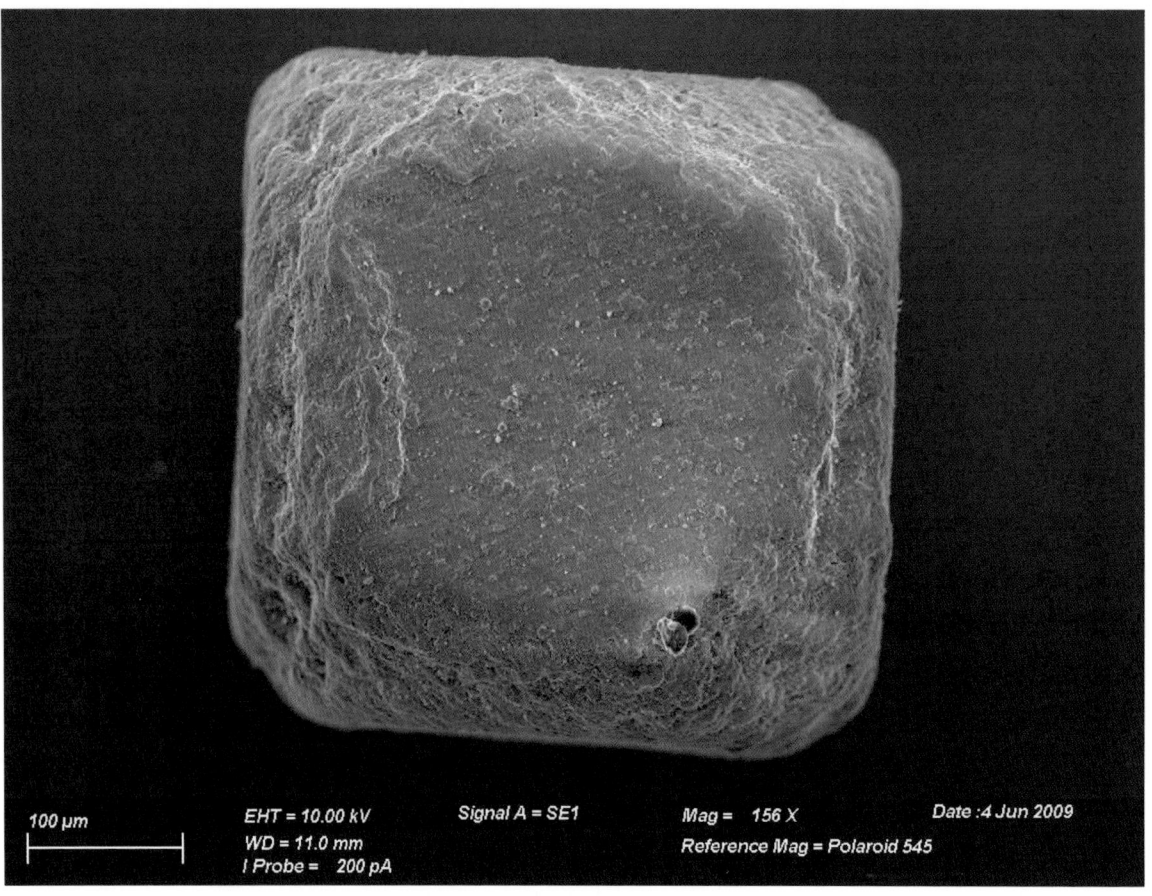

In two dimensions, the triangle is the strongest single polygon.

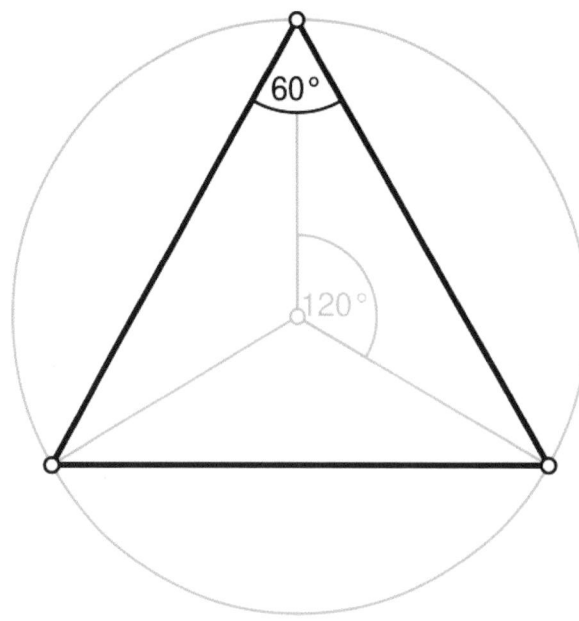

Pyramids are a good example of the strength of triangles on a square base. Remember that 4/3 is the ratio for the perfect musical fourth.

This strong but flimsy looking tetrahedron shaped kite is a good example of triangular power:

Geodesic domes are another good example of the power of triangles. Most geodesic domes are based on the Archimedean solid, the icosahedron, which is made entirely from them.

To make a standard dome, each triangle side of the icosahedron is divided by adding smaller triangles inside them. This always results in a regular number amount of smaller triangles.

3 triangles:

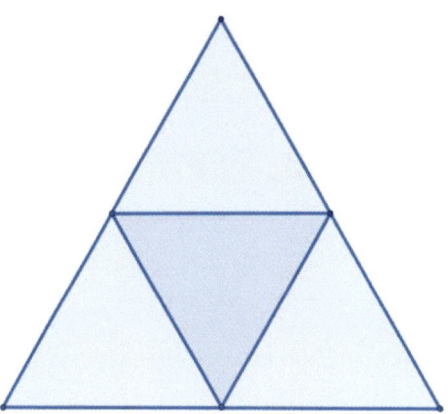

9 triangles:

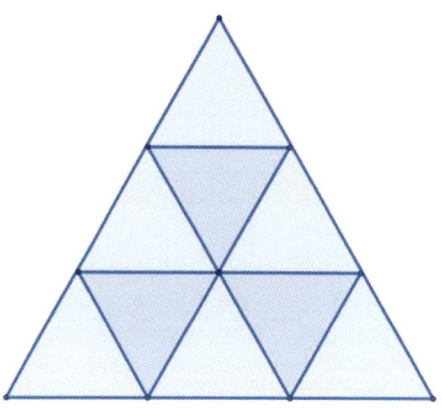

16 triangles:

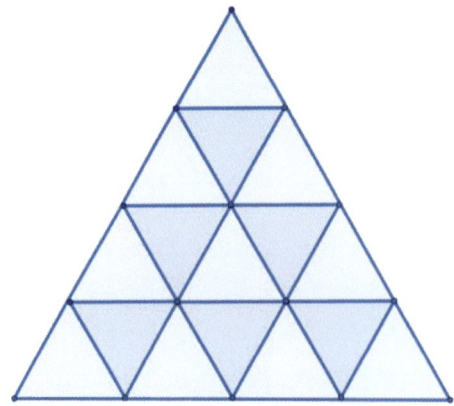

I don't know all of the details involved in dome building, but I know that the lengths of some struts are adjusted to make these flat triangles slightly curved. This then makes a geodesic sphere which always has a regular number amount of triangles. This sphere can then be cut in two at various points to make a flatter or taller dome.

I lived in a large one made from very thin aluminum struts about a meter long with a tent-like cover over it. The highest point was much higher than any house I have lived in. These struts were so weak that I could break the spare struts on my knee, but when they were joined in the dome I could climb-up the inside without it even bending much. Somehow the geometry added a lot of strength to the weak parts it was made from. I remember that the person who built it told me that when he added the last strut, the whole dome rang out a clear tone.

Geodesic domes are usually made from only triangles, but I have seen ones based on other Platonic or Archimedean solids. Domes based on the truncated icosahedron made from pentagons and hexagons together for example are also quite common. Remember that the perfect just intonation minor third has a ratio of 6/5.

Most soccer balls are truncated icosahedrons, many players swear that this is and always will be the best shape for a soccer ball to be.

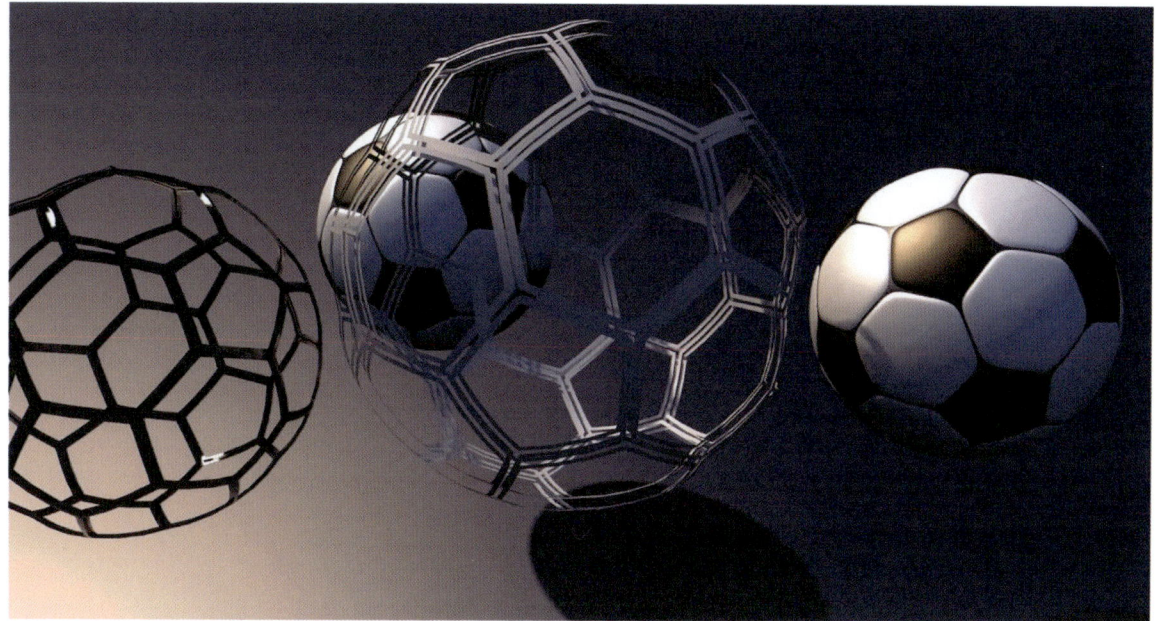

A fullerene is an allotrope of carbon in the form of a hollow sphere, ellipsoid, tube, and many other shapes. They were named this because some of them resemble geodesic domes which were invented by Buckminster Fuller.

Buckminsterfullerene is a type of fullerene with the formula C60 that was discovered in 1985. It is a tiny truncated icosahedron (12 pentagons and 20 hexagons), with a carbon atom at each vertex of each polygon and a bond along each polygon edge. It is basically a very small soccer ball.

C60 is a remarkably stable compound.

C60 has been characterized as a "free radical sponge" with an antioxidant efficacy several hundred-fold higher than conventional antioxidants. It has shown promise in many medical treatments for various ailments from repairing bones and cartilage to healing cancer and various other illnesses.

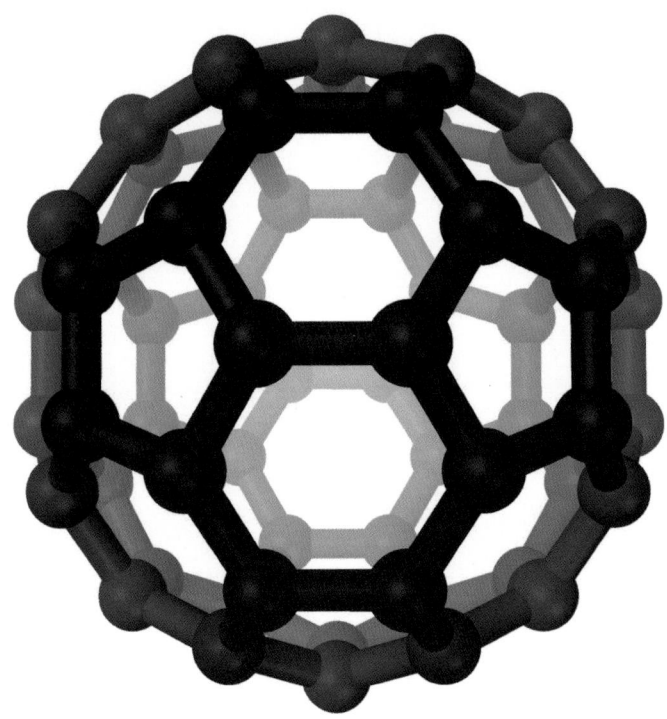

Cylindrical fullerenes are also called carbon nanotubes. These are allotropes of carbon with a cylindrical nanostructure that are usually made from only hexagons which when joined together make a single cylindrical molecule that can be any length. The chemical bonding of nanotubes are similar to those of graphite and stronger than those found in diamonds. These cylindrical carbon molecules have unusual properties, which are valuable for nanotechnology, electronics, optics and other fields. Because of its extreme strength and stiffness, nanotubes have been constructed with length-to-diameter ratio of up to 132,000,000. This is much larger than for any other material.

Carbon nanotube:

A scanning tunneling microscopy image of single-walled carbon nanotube:

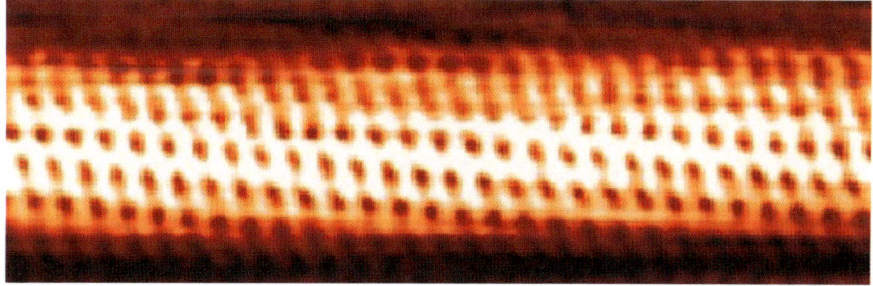

When I look at the above image, I see triangles forming hexagons, and not hexagons on their own. This does make sense and would be stronger that just hexagons. I am no scientist though, so maybe I just don't understand the image...

As you can see, all of these things are formed around regular numbers. Using other numbers does not work as well at all. Here are some examples of heptahedra (a heptahedron has 7 sides).

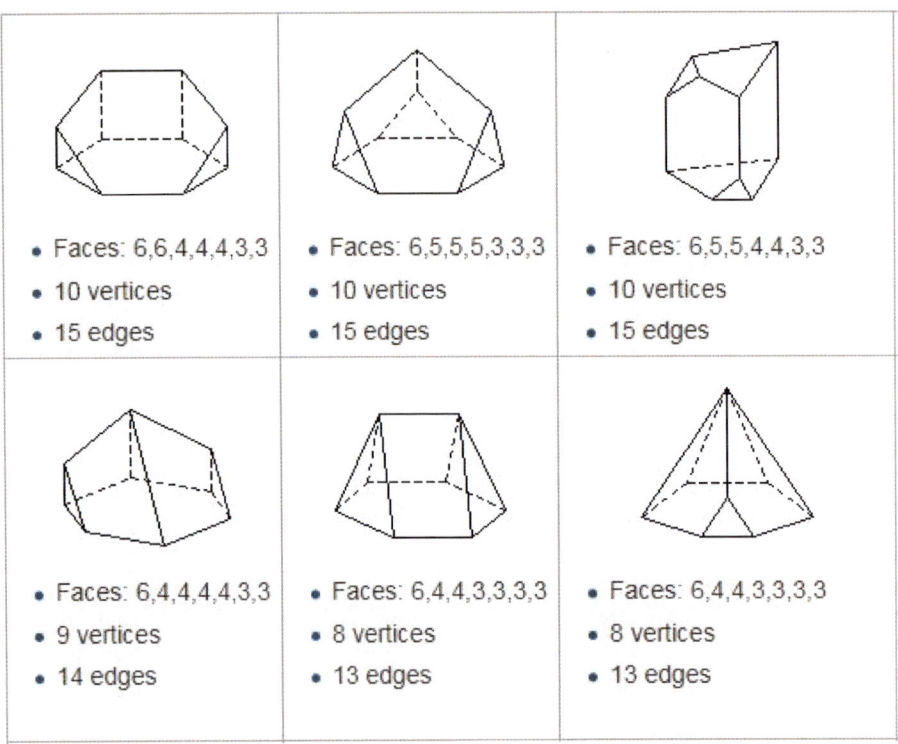

- Faces: 6,6,4,4,4,3,3
- 10 vertices
- 15 edges

- Faces: 6,5,5,5,3,3,3
- 10 vertices
- 15 edges

- Faces: 6,5,5,4,4,3,3
- 10 vertices
- 15 edges

- Faces: 6,4,4,4,4,3,3
- 9 vertices
- 14 edges

- Faces: 6,4,4,3,3,3,3
- 8 vertices
- 13 edges

- Faces: 6,4,4,3,3,3,3
- 8 vertices
- 13 edges

Prisms are made from two of the same polygons with squares joining them together. A prism made from 7 sided polygons is quite regular, but you can't stack them well to make a wall because of the uneven amount of sides.

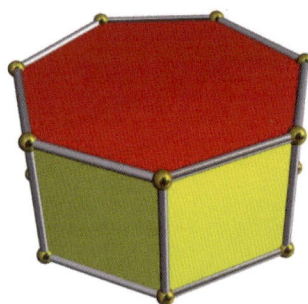

If you make one from two 5 sided pentagons then the whole prism will have 7 sides including the top and bottom, but this also will not stack well.

So, objects based on 7 will be much easier to break than the more regular Platonic and Archimedean solids, geodesic domes, carbon nanotubes, and pyramids will be. The irregularity of these shapes creates many weak spots where it will easily fail, and stacking them in a stronger way like you can with cubes is usually impossible because of the odd numbers of sides they have. Things don't get any better with 11, 13, 14 etc sided polyhedrons.

So, regular number mathematics seems to represent strength in physical objects as well as musical harmony. While "irregular" numbers like 7, 11, 13 and 14 do not.

Ancient Mathematics

Many people wonder why the mathematics of certain ancient cultures is often very geometrical and musical, if you use the numbers as degrees or Hz frequencies. The quick explanation for this is that they used regular numbers to simplify equations, and regular numbers are what also you use to make geometrical objects and 5-limit music scales.

Mayan Long Count

The Mayans did not use base 10 mathematics as we do today. Instead the "Long Count" days were tallied in a modified base 20 scheme to work better with 360 day years than normal base 20 does. The Mayans only needed symbols for the numbers 1 to 20 to do this mathematics.

In a pure base 20 scheme, 0.0.0.1.5 is equal to 25 and 0.0.0.2.0 is equal to 40. The Long Count is not pure base 20, however, since the second digit from the right (and only that digit) rolls over to zero when it reaches 18 instead of 20. Thus 0.0.1.0.0 does not represent 400 days, but rather only 360 days.

The Mayan calendar is quite simple. It starts with 1 day (Kin), then 20 days (Winal), then 20 is multiplied by 18 instead of 20 to give 360 days (Tun). After that each number is multiplied by 20 to get the next one.

Mayan long count calendar			
Baktun	20 Katuns	20 x 7200 =	144000 days
Katun	20 Tuns	20 x 360 =	7200 days
Tun	18 Winals	18 x 20 =	360 days
Winal	20 Kins		20 days
Kin			1 day

The amounts of days are all regular numbers, you can tell by their prime factors (2, 3 and 5).

Mayan long count prime factors			
Baktun	20 Katuns	$2^7 \times 3^2 \times 5^3 =$	144000 days
Katun	20 Tuns	$2^5 \times 3^2 \times 5^2 =$	7200 days
Tun	18 Winals	$2^3 \times 3^2 \times 5 =$	360 days
Winal	20 Kins	$2^2 \times 5 =$	20 days
Kin			1 day

Although the following numbers are not part of the Long Count, the Mayans had names for larger time spans. This is an extension of the 5 positions in the Long Count calendar; each number is just multiplied by 20 to get the next one. Because $2^2 \times 5 = 20$, you can tell that these large amounts of days are also regular numbers.

Longer time spans				
Alautun	20 Kinchiltuns	20 x 1152000000 =	23040000000 days	approx. 63 million years
Kinchiltun	20 Calabtuns	20 x 57600000 =	1152000000 days	approx. 3 million years
Calabtun	20 Pictuns	20 x 2880000 =	57600000 days	approx. 158,000 years
Pictun	20 Baktuns	20 x 144000 =	2880000 days	approx. 7885 years

As you can see, this math enabled the Mayans to measure time very far into the future, in a very simple way. The resulting numbers were large but were still regular numbers, so they can be used for more calculations that will also be simplified. This is a good example of the mathematical power of regular numbers.

Regular number based mathematics is always musical. In this case multiplying a frequency by 18 raises it by 4 octaves and a 9/8 major whole tone, and multiplying a frequency by 20 raises of by 4 octaves and a 5/4 major third. So the spaces between these time spans are very harmonious.

The number 144000 (20 Katuns in the Mayan long count) has religious significance for Christians. It appears three times in the Book of Revelation.

Revelation 7:3–8:

"Do not harm the earth or the sea or the trees, until after we have sealed the servants of God on their foreheads."

"And I heard the number of the sealed, a hundred and forty-four thousand, sealed from every tribe of the sons of Israel:"

12000 from the tribe of Judah were sealed,
12000 from the tribe of Reuben,
12000 from the tribe of Gad,
12000 from the tribe of Asher,
12000 from the tribe of Naphtali,
12000 from the tribe of Manasseh,
12000 from the tribe of Simeon,
12000 from the tribe of Levi,
12000 from the tribe of Issachar,
12000 from the tribe of Zebulun,
12000 from the tribe of Joseph,
12000 from the tribe of Benjamin were sealed.

(12000 x 12 = 144000)

Revelation 14:1:

"Then I looked, and behold, on Mount Zion stood the Lamb, and with him 144000 who had his name and his Father's name written on their foreheads".

Revelation 14:3–5:

"And they were singing a new song before the throne and before the four living creatures and before the elders. No one could learn that song except the 144000 who had been redeemed from the earth. For it is these who have not defiled themselves with women, for they are virgins. It is these who follow the Lamb wherever he goes. These have been redeemed from mankind as first fruits for God and the Lamb, and in their mouth no lie was found, for they are blameless."

Page from an illuminated manuscript painted by an unknown artist, depicting the 144000 from Revelation 7:

Jehovah's Witnesses believe that exactly 144000 faithful Christians will be resurrected to serve as priest kings to lord over the little people along with God and Jesus... The Church of Jesus Christ of Latter-day Saints also believe some weird stuff about 144000 high priests. The Christian Skoptsy sect in Russia believed that the Messiah would come when there were 144000 Skoptsy believers. The Unification Church founded by Reverend Sun Myung Moon believes the 144000 represents the total number of saints whom Christ must find to save the world from Satan.

I am not sure what these Christians were up to, but $2^7 \times 3^2 \times 5^3 = 144000$ making it a regular number.

Ancient Indian mathematics is very complex, so I am not sure how they made the Yugas. Regular numbers also seem to have been involved while 7, 11, and 13 were avoided, though. You can tell this by the 12's, 24's, 36's and 48's with added 0's that you can see in them. To add a 0 to a number with 5-limit primes, just multiply it by 5 and by 2. So, 36 x 5 x 2 = 360, and 360 x 5 x 2 = 3600. Hence, if 36 is regular so is 360, 3600, 36000 etc. And so you can be sure the all of the numbers below are regular numbers because 12, 24, 36 and 48 are all regular numbers with added 0's.

The Yugas

Yugas		Years
Satya-Yuga	Golden age = all good	4800 years
Treta-Yuga	3/4 good 1/4 dark	3600 years
Dwapara-Yuga	50 / 50 good and dark	2400 years
Kali-Yuga	Dark Age = 1/4 good 3/4 dark	1200 years

All 4 Yugas (4800 + 3600 + 2400 + 1200) = 12000 years

4 Yugas	One arc of precession	12000 years
2 x 4 Yugas	Full precession	24000 years

12, 24 and 48 are all octaves of G in Ptolemy's intense diatonic scale with 24 Hz as reference pitch, while 36 is an octave of D. So musically octaves and a perfect 3/2 fifth exist in the Yugas.

The Babylonians inherited their mathematics from the Sumerians, who were the first known civilization on Earth. From them it seems to have filtered down through various cultures, all the way to our modern culture where we still use some of it to measure time, distance, and angles.

Their influence seems to have spread far and wide, as you will find concepts specific numbers first seen in the Sumerian and Babylonian cultures in the mythologies of India, China, Babylon, Greece, Israel, and Europe.

How the they came upon this math that is so advanced, so long ago, is a mystery. We know what they did mathematically because they left many stone tablets, and math is pretty logical to understand. But the tablets describing their history can be translated in many ways. Some translations say that they said they learned it from the Gods who came from the east. Some people say this means aliens, but I think it is more likely that another civilization just like ours existed long ago in the east from where Sumer and Babylon were. Or, maybe they just discovered it and attributed it to the Gods of inspiration. Nobody really knows.

Base 60

The Babylonians used a Sexagesimal (base 60) numeral system with sixty as its base. The number 60, a superior highly composite number, has twelve factors, namely 1, 2, 3, 4, 5, 6, 10, 12, 15, 20, 30, and 60, of which 2, 3, and 5 are prime numbers. With so many factors, many fractions involving sexagesimal numbers are simplified. For example, one hour can be divided evenly into sections of 30 minutes, 20 minutes, 15 minutes, 12 minutes, 10 minutes, 6 minutes, 5 minutes, 4 minutes, 3 minutes, 2 minutes, and 1 minute.

60 is also the smallest number that is divisible by every number from 1 to 6; that is, it is the lowest common multiple of 1, 2, 3, 4, 5, and 6.

The Babylonians only had symbols for the numbers 1 to 60, with 60 being written as a large 1. To write larger numbers they would multiply these numbers by powers of 60.

The first few powers of 60 are: 60, 3600 and 216000

60 x 1 = 60
60 x 60 = 3600
60 x 60 x 60 = 216000

They did not stop on 216000 though, and did extend this to larger powers of 60 when needed. When they wrote a number they would use this as a place value system:

Base 60 place value system			
216000 times any of the numbers from 1 to 60	3600 times any of the numbers from 1 to 60	60 times any of the numbers from 1 to 60	Any of the numbers from 1 to 60

The number 72 for example would be written as 10 and a 2 in the far right column, and a 1 in the second from the right column. This would equal 60 + 10 + 2 = 72.

This way of doing mathematics was used to define the second as 1/86400 of a day. So, one 24 hour day has 86400 seconds in it (24 x 3600 = 86400), and a 12 hour day has 43200 seconds (12 x 3600 = 43200). To write 43200 the Babylonians would simply place a 12 in the third column from the right. For 86400 they would place a 24 in the same place.

The standard multiplication tables of the Babylonian sexagesimal system were generated from these 30 numbers and their reciprocals (inverse number). I don't fully understand these reciprocals written in base 60 using modern numbers (like a studied mathematician might), but it is the main 30 numbers on the left of the chart that are of most interest to me, because they are all the regular numbers up to 60. Remember that regular numbers evenly divide powers of 60. So, they are really good to use with base 60 mathematics.

Regular numbers	Reciprocals
2	30
3	20
4	15
5	12
6	10
8	7,30
9	6,40
10	6
12	5
15	4
16	3,45
18	3,20
20	3
24	2,30
25	2,24
27	2,13,20
30	2
32	1,52,30
36	1,40
40	1,30
45	1,20
48	1,15
50	1,12
54	1,6,40
60	1

The table excludes 7, 11, 13, 14, 17 etc because using them in these calculations resulted in irrational numbers. They did know how to compute these numbers when they had to, but they used other more complex tables and sums that often estimated things and were not always 100% accurate.

This was very useful in the amazing method of multiplication by reciprocal used by Sumerians and Babylonians. Numbers were never actually divided by another number, they were instead multiplied by the reciprocal (or inverse) of the other number. In this system, the expression "60 divided by 10" becomes "60 multiplied by one-tenth." [60 / 10 = 6] is the same as [60 x 0.1 = 6]

Base 60 math works very well together with the 360 divisions of a circle and 360 day years because all divisors of 60 are also divisors of 360.

Divisors of 60 and 360																								
60		2	3	4	5	6			10	12	15		20		30			60						
360	1	2	3	4	5	6	8	9	10	12	15	18	20	24	30	36	40	45	60	72	90	120	180	360

This is easy to visualize with triangles inside a hexagon. A triangle contains 60 degrees in one corner, so 6 of these will fill a 360 degree circle perfectly (60 x 6 = 360).

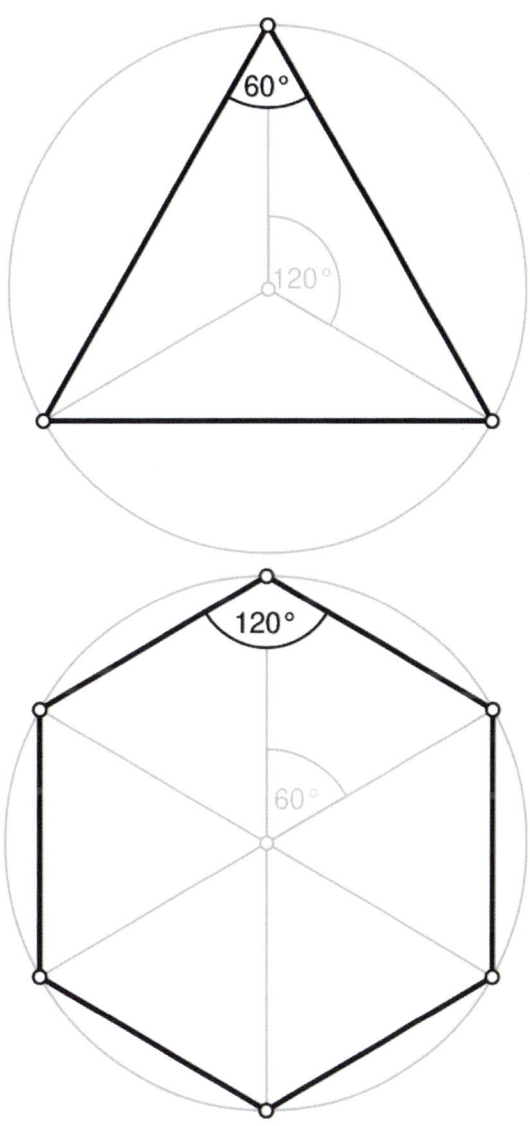

Kings list

This is an ancient list that nobody really understands. It seems to show a list of kings that ruled before the "great flood". Some say this is the same flood spoken of in the Bible. What does not make sense is that these kings seemed to live for quite a few thousand years each.

If you look at this list you can see that 28800, 36000 and 43200 (also found in many other cultures' texts) are very easy to reach using base 60 math. This is mainly because 3600 is a power of 60, so 3600 is like a default setting in base 60 mathematics, all you need to do is to multiply it by 8, 10 or 12.

Ruler	Measure		Length of reign
Kingship descends from heaven			
Alulim	8 Sars	8 x 3600 =	28800 years
Alalngar	10 Sars	10 x 3600 =	36000 years
En-men-lu-ana	12 Sars	12 x 3600 =	43200 years
En-man-gal-ana	8 Sars	8 x 3600 =	28800 years
Dumuzid	10 Sars	10 x 3600 =	36000 years
En-sipad-zid-ana	8 Sars	8 x 3600 =	28800 years
En-man-dur-ana	5 Sars and 5 Ners	5 x 3600 + 5 x 600 =	21000 years
Ubara-Tutu	5 Sars and 1 Ner	5 x 3600 + 600 =	18600 years
Great Flood			

28800, 36000, and 43200 make a perfect 4-5-6 major chord, only it is above the human hearing range.

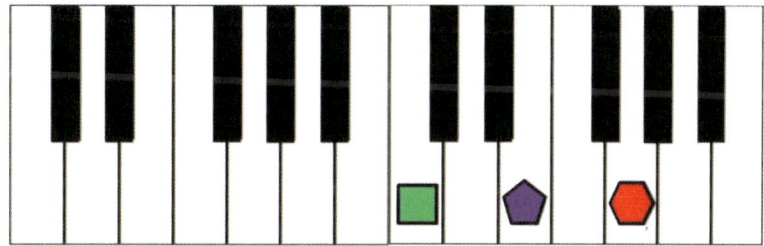

The two kings at the bottom make strange intervals, and have the numbers 21000 and 18600 which are not regular numbers. This is because addition and not only multiplication was used in those sums, excluding the additions 5 x 3600 = 18000 and 5 x 600 = 3000, which are both regular numbers. The interval between them is a 4/3 perfect fourth.

So, the regular number based kings list is a very musical list indeed.

NUMBERS OF THE GODS

I have "researched" the descriptions and stories of these deities shamelessly from Wikipedia and a few other websites. I have kept them short, so if you want to know more about each deity, just search for their names online.

What I have added that is harder to find is the number associated with each deity. Apart from 2 of them at the end of this chapter which I have clearly marked as vague, the numbers I have provided are 100 % correct. I have checked many sources and the numbers for the most important deities are literally written in stone (stone tablets).

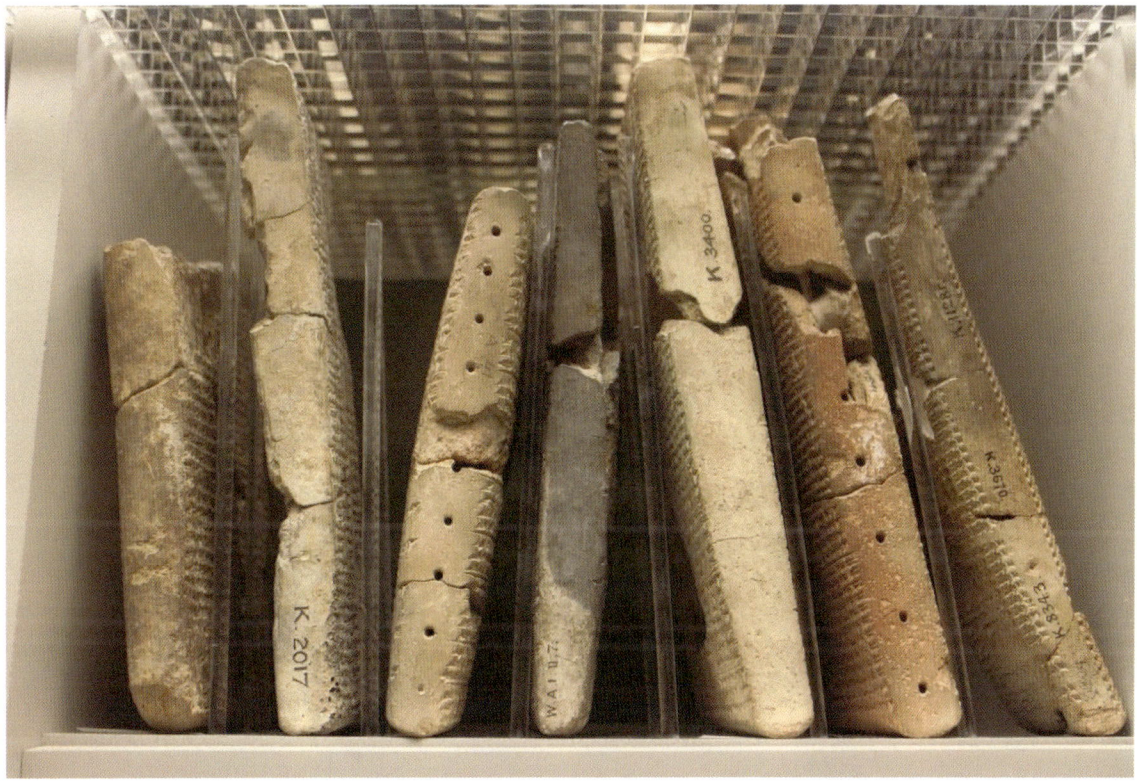

It seems that more numbered deities than I can find also exist, you can tell by the gaps in my charts. Also many deities did not seem to have numbers at all, so they are also not highlighted here. This includes the primordial deities Abzu, Anshar and Kishar, Ki, Nammu and Tiamat, who are only mentioned briefly.

The hierarchy of importance among the numbered deities of Babylonian (that I have found) runs in additions of five from 5 to 60. Females had numbers ending in 5, and were assigned to the "feminine" side because only males could rule. Each male God's female consort has the same number minus 5 i.e. Antu (55) was the consort of Anu (60); Ninlil (45) was the consort of Enlil (50), and so on. These Gods had fights and affairs just like humans do, so things were not always cut and dry, but generally the consorts were meant to be the Gods' numbers minus 5.

Male Gods		Female consorts	
60	Anu	55	Antu
50	Enlil	45	Ninlil
40	Enki	35	Ninki
30	Sin	25	Ningal
20	Shamash	15	Ishtar
10	Adad	5	Ninhursag

The spiritual politics of post-Sumerian Mesopotamia resulted in the altering of the names or figures assuming the titles, but not the roles themselves (which were mathematically fixed). With the passing of each generation, the successive characters moved up in their positions, leaving us with the system illustrated here, the most "updated" post-Sumerian version of the Pantheon.

Note: In the piano examples in this chapter, I have placed the notes as if the 0's are not there. This makes more sense when using these numbers as Hz frequencies. 40 – 50 – 60 for example is placed on the notes for 4 – 5 – 6, higher octaves of 4, 5 and 6 are 256, 320 and 384 Hz, these are the notes for the C major chord in Ptolemy's intense diatonic scale as it is in this book. If I had used 40 Hz, 50 Hz and 60 Hz it would have been an E major chord in the same scale, the exact same chord only in a different key that includes a black key. Using the numbers without the 0's however fits better into the white keys of the keyboard making more sense, so I have used them instead.

HIGHEST TRIAD

Enki (40), Enlil (50), and Anu (60) are the three highest Gods.

Notice that they define the 4 – 5 – 6 major chord, the same chord found in the kings list only in a lower key.

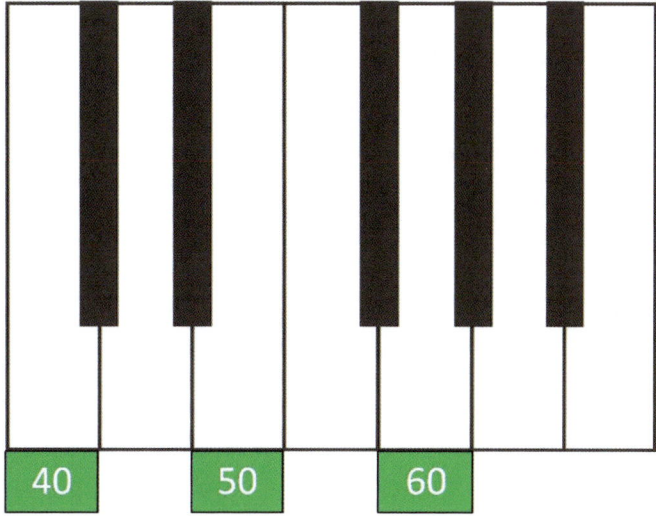

All of the lower deities are the descendants of Enki (40), Enlil (50), and Anu (60.

The designations of the Supernal Trinity, 40, 50 and 60, were "master numbers." The Babylonian fractional (lower) designations are assumed by the "younger pantheon"; 40 and 50 are not divisors of 60, showing that Enlil and Enki are more important than the other less important, but still important deities which are usually divisors or fractions of 60 / Anu the supreme God i.e. 40 and 50 stand alone as 60 does, yet are not as superior to 60.

Deities	Regular numbers	Divisors of 60	Fraction of 60
	1	1	1/60
	2	2	1/30
	3	3	1/20
	4	4	1/15
	5	5	1/12
	6	6	1/10
	8		
	9		
	10	10	1/6
	12	12	1/5
	15	15	1/4
	16		
	18		
	20	20	1/3
	24		
	25		
	27		
	30	30	1/2
	32		
	36		
Enki	40		
	45		
	48		
Enlil	50		
	54		
Anu	60	60	1/1

Anu (60)

Written as a large 1, Anu (60) is the supreme God; the divine personification of the sky. He was believed to be the highest source of all authority, for all the other Gods and mortal rulers, and was described as "the one who contains the entire universe". This makes sense as their base 60 mathematical system created from 60.

Note that 60 Hz is a very good reference pitch for just intonation scales with low decimal Hz frequencies.

Anu and the Sumerian creation myth: The prologue to the epic poem Gilgamesh, Enkidu, and the Netherworld briefly describes the process of creation. At first there was only Nammu, the primeval sea. Nammu then gives birth to An (the Sumerian name for Anu), seen as the sky, and Ki, seen as the Earth. An and Ki mate with Ki giving birth to Enlil, the God of the wind. Enlil separates An from Ki and declares Earth as his domain, while An keeps the sky.

Enlil (50)

Enlil became later known as Elil and is associated with wind, air, Earth, and storms. He was first seen as the chief deity of the Sumerian pantheon, but went on to be worshipped by the Akkadians, Babylonians, Assyrians, and Hurrians. His "Ekur temple" in the city of Nippur was regarded as the "mooring-rope" of heaven and Earth. Enlil is also, at times, referred to in Sumerian texts as Nunamnir. According to a Sumerian hymn, he was so holy that the other Gods couldn't even look at him.

Enlil plays a vital role in the Sumerian creation myth, separating An (heaven) from Ki (Earth), making the Earth habitable for humans. In the Sumerian Flood myth, Enlil rewards Ziusudra with immortality for having survived the flood and, in the Babylonian flood myth, Enlil is the cause of the flood himself, having sent the flood to exterminate the human race who made too much noise and prevented him from sleeping (disturbed the peace!).

It is interesting to note that the Greek God Pan also disliked being woken from his sleep i.e. disharmony is often shunned by Gods throughout history.

Enlil (50) and Ninlil (45) conceived the moon-God Nanna (30), as well as the Underworld deities Nergal, Ninazu, and Enbilulu.

Enki (40)

Enki 40 is the Sumerian God of water, knowledge, mischief, crafts, and creation, and later became known as Ea in Akkadian and Babylonian mythology. The planet Mercury was also identified with Enki.

Enki was the keeper of the divine powers called Me, the gifts of civilization. He is often shown with the horned crown of divinity. He was considered the master shaper of the world, the God of wisdom and of all magic.

Astral triad

These three deities, Sin, the Moon God (30), Shamash, God of the Sun (20) and Ishtar, the Goddess of Venus (15) formed a secondary astral triad. 30, 20, and 15 are all divisors or fractions of 60. This shows that they are less important than the 3 main Gods which are not divisors of 60, but still important enough to get numbers closely related to 60. The fact that these 3 deities were given the highest regular numbers that are also divisors of 60 must mean something.

Note: musically 15, 20 and 30 are lower octaves of 60 and 40.

Deities	Regular numbers	Divisors of 60	Fraction of 60
	1	1	1/60
	2	2	1/30
	3	3	1/20
	4	4	1/15
	5	5	1/12
	6	6	1/10
	8		
	9		
	10	10	1/6
	12	12	1/5
Ishtar	15	15	1/4
	16		
	18		
Shamash	20	20	1/3
	24		
	25		
	27		
Sin	30	30	1/2
	32		
	36		
Enki	40		
	45		
	48		
Enlil	50		
	54		
Anu	60	60	1/1

Musically this triad is not a major chord, it contains an octave, a perfect fourth and a perfect fifth. Still a very musical triad.

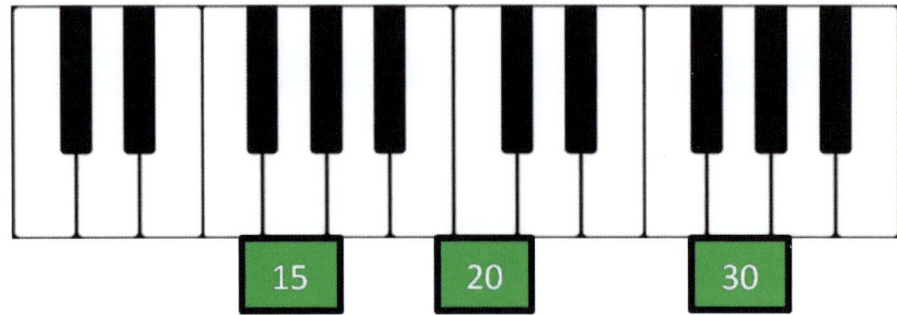

The Sun (20) and Moon (30) are obviously the most important to us on Earth, which is why they have numbers that are so close to 60. Venus (15) must be the next most important because it has a smaller number, this makes sense considering what the orbit of Venus looks like from Earth, and the fact that Venus is the morning and evening star.

Venus orbit:

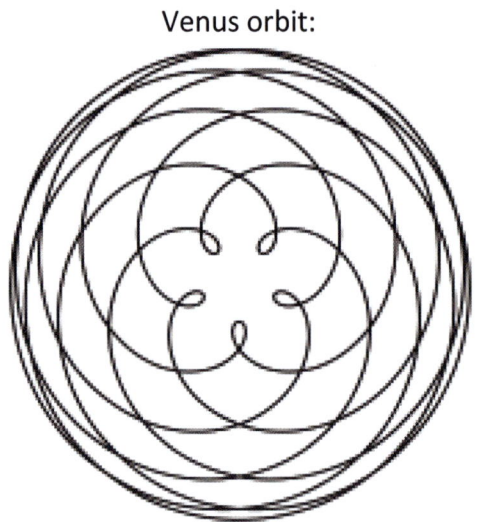

Sin (30) (Moon)

Sin 30 (1/2 of 60). Sin is the moon God, which is interesting because 30 days = lunar month. Sin is also known as Nanna, the son of Enlil and Ninlil (as mentioned above). When the area of Ur exercised supremacy over the Euphrates valley (between 2600 and 2400 BC), Sīn/Nanna was considered the supreme god, appointed as the "father" or "head" of the Gods - "creator of all things". His concubine was Ningal (25).

Shamash (20) (Sun)

Shamash 20 (1/3 of 60); the Sun - judge of the Gods and men - was the son of Sin. As the solar deity he exercised the power of light over darkness and evil, and became known as the "God of justice and equity". At night, he became judge of the underworld. Shamash was also one of the Gods of oracles and of divination.

Ishtar (15) (Venus)

Ishtar 15 (1/4 of 60); She was the epitome of the feminine, and considered the most important female deity of ancient Mesopotamia throughout time.

According to Sumerian tradition only males could give commands. Ishtar (also known as Innana) refused to accept this ruling and constantly plotted to gain power.

She is sometimes seen as the daughter of the sky God An, and then sometimes as his wife. In other myths she is the daughter of Nanna or Enlil. Ishtar was seen as the Goddess of dates, wool, meat, and grain, as well as the Goddess of rain and thunderstorms (which led to her association with An, the sky God). She was also a fertility figure, and sometimes referred to as the Lady of the Date Clusters.

Although seen as a fertility figure, she was often surrounded in myth by death and disaster, so also became seen as a goddess of contradictory connotations and forces.

Associated with Venus, Ishtar was the protectress of prostitutes and the patroness of the alehouse. Her popularity was universal in the ancient Middle East, and in many cases she played the role of numerous local Goddesses. In later myth she became known as Queen of the Universe, taking on the powers of Anu, Enlil, and Enki.

7 MOST POWERFUL DEITIES

The above 6 deities were part of a group known as the 7 most powerful and important deities of Sumer and were probably in the Anunnaki. Although certain deities are described as members of the Anunnaki, no complete list of the names of all the Anunnaki has survived. In the earliest texts, the term is applied to the most powerful and important deities in the Sumerian pantheon; the descendants of the sky-God, Anu.

This group of deities probably included the "7 most powerful deities"

7 most powerful deities	
Anu	60
Enlil	50
Enki	40
Sin	30
Shamash	20
Ishtar	15
Ninhursag	5

The only deity here I have not described yet is Ninhursag.

Ninhursag (5)

As you know, 5 is a very important prime number.

Also known as Damgalnuna or Ninmah, Ninhursag was seen as the ancient Sumerian mother Goddess of the mountains, and one of the seven great deities of Sumer. She is principally a fertility goddess. Sometimes her hair is seen in an omega shape, and other times she wears a horned headdress and tiered skirt, often with bow cases on her shoulders. She also carries a mace or baton with an omega motif (or something similar), and a lion cub on a leash can be seen by her side.

Deities	Regular numbers	Divisors of 60	Fraction of 60
	1	1	1 / 60
	2	2	1 / 30
	3	3	1 / 20
	4	4	1 / 15
Ninhursag	5	5	1 / 12
	6	6	1 / 10
	8		
	9		
	10	10	1 / 6
	12	12	1 / 5
Ishtar	15	15	1 / 4
	16		
	18		
Shamash	20	20	1 / 3
	24		
	25		
	27		
Sin	30	30	1 / 2
	32		
	36		
Enki	40		
	45		
	48		
Enlil	50		
	54		
Anu	60	60	1 / 1

These 7 most important deities were given the most musical numbers of all, the 4 – 5 – 6 major chord with 4 lower octaves of those 3 notes including 3 – 4 – 5 and 2 – 3 – 5 major chords. The low 5 provides the perfect bass note at the root note of the chord, exactly where you would put a bassline in music based on that chord. This set of notes really sounds like the choirs of heaven itself.

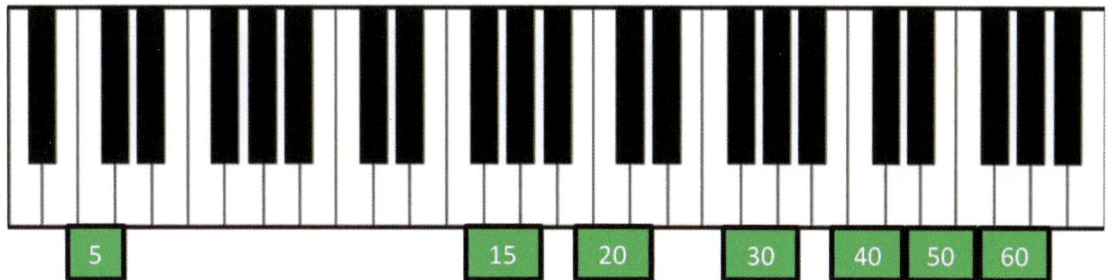

7 PLANETARY DEITIES

Each of these 7 planets had a deity assigned to them. Sin, Shamash and Ishtar have already been described. Marduk, Ninurta, Nergal and Nabu have their own interesting stories.

7 Planetary Deities	
Sin	Moon
Shamash	Sun
Ishtar	Venus
Ninurta	Saturn
Marduk	Jupiter
Nergal	Mars
Nabu	Mercury

The number 50 not only represented Enlil, but also his position as chief of all the activities on Earth. After the Deluge / flood, when the younger Gods challenged the authority of the older Gods for leadership, Enlil's military aide Ninurta assumed the title of 'fifty' and claimed leadership; the role which Enlil apparently aborted.

With the eclipse of the senior Gods after the Deluge, there was a scramble for power among the younger Gods (with Ishtar seemingly always involved). This brought much disorder to the nations of the Middle East.

The numbers below are associated with the deities, which were then assigned planets. So I don't think these numbers are the numbers of the planets themselves. Unlike the above deities, I am also not sure if 12 and 14 are actually correct. This is because some less important deities were not mentioned as often, and so details of them are harder to come by than those for the major deities.

Musically you already know that the Moon (30), the Sun (20) and Venus (15) make a good triad. 50 is good too. 14 is the devil's interval which is great for the God of the underworld. The not so important Nabu and his 12 make a minor third in C which kind of clashes with the major chord in C for the 7 Gods that decree. Keep in mind that 12 and 14 may not be correct.

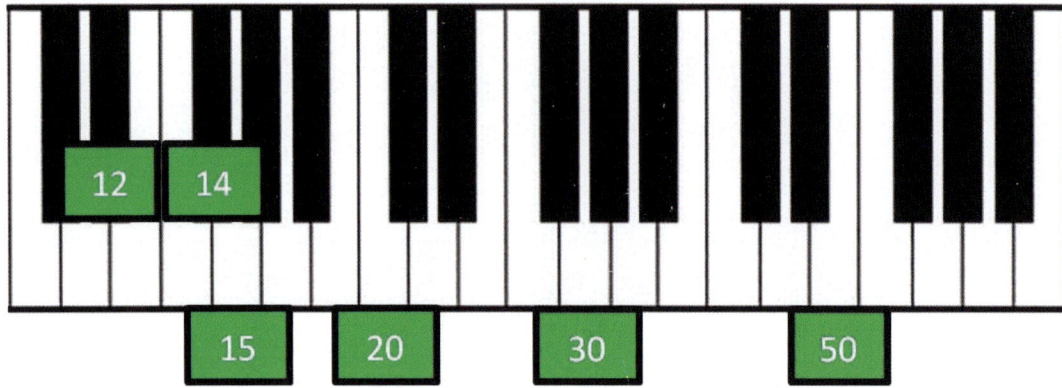

Ninurta (50) (Saturn)

Second only to the Goddess Inanna / ishtar, Ninurta played a combined role as both a warrior deity and an agricultural deity. He claimed leadership after Enlil "resigned" from his position.

Marduk (50) (Jupiter)

Marduk 50; Jupiter; also known as Bel or the biblical "Baal" - he "inherited" all the powers of the other Gods. Besides Ninurta being in the role of leadership, Marduk, the eldest son of Enki, also assumed the title of "fifty" as he proclaimed himself chief of Babylon, even though he was was unranked in the pantheon at first.

A late-generation God from ancient Mesopotamia, Marduk was associated with the divine wind weapon, Imhullu. His consort was the goddess Sarpanit, and he was seen as the son of Ea / Enki) and Damkina - and the heir of Anu. Enki and Enlil passed their powers and attributes over to Marduk.

Under his reign humans were made to bear the burdens of life so the Gods could be at leisure.

Nergal (14?) (Mars)

Nergal is not actually placed on the original post-primordial pantheon. In a text I found, it says that Nergal is 14 - although I have not found more proof of this. However, Nergal was a God of the underworld, and associated with Mars, so it is likely that they would have given him a number like 7, 11, 13 or 14.
It makes sense that the Martian God would be "out-of-tune" with the other Gods.

Nergal is a son of Enlil and Ninlil, along with Nanna and Ninurta. Portrayed in hymns and myths as a God of war and pestilence, Nergal seems to represent the sun of noontime and of the summer solstice that brings destruction; high summer being the "dead season" in the Mesopotamian annual cycle. Over time Nergal developed from a war God to a God of the Underworld. This occurred when Enlil and Ninlil *gave* him the underworld.

Nabu (12?) (Mercury)

I have seen Nabu as 12 in a single piece of text, but nowhere else. Nabu was identified as the son of the God Marduk. His wife was the Akkadian Goddess Tashmet.

He was the patron God of scribes, literacy, and wisdom; the inventor of writing, a divine scribe, and the patron God of the rational arts. Due to his role as an oracle, Nabu was associated with the Mesopotamian Moon God, Sin.

Nabu wore a horned cap and stood with his hands clasped in the ancient gesture of priesthood. He rode on a winged dragon known as Sirrush that originally belonged to his father, Marduk. In Babylonian astrology, Nabu was also identified with the planet Mercury.

OTHER DEITIES

Adad (10)

Adad (also known as Hadad) was the storm and rain God in the Northwest Semitic and ancient Mesopotamian religions. The bull was his symbolic animal. He was of less importance, probably because storms and rains were scarce in Sumer and agriculture there depended on irrigation. The gods Enlil and Ninurta also had storm God features, so he sometimes appeared as their "assistant" or "companion".

In other texts, Adad (also known as Iskur) is sometimes seen as the son of the moon God Nanna / Sin fathered by Ningal, and brother of Utu / Shamash and Inanna / Ishtar. He is also sometimes described as the son of Enlil. Shamash and Adad, in combination, became the Gods of oracles and divination.

10 is a regular number and a divisor of 60. It seems like quite an important number for an unimportant God. Maybe these Gods existed and got their numbers long before Sumer? Perhaps Adad was much more important in a long ago rainier time?

Ningal (25)

Ningal was a Goddess of reeds in Sumerian mythology, the daughter of Enki and Ningikurga, and the consort of the moon God Nanna / Sin (30) with whom she conceived the sun God Utu / Shamash and his sister, Inanna. 25 is a regular number but not a divisor of 60.

Ninki (35)

Ninki also known as "lady of the earth" was the spouse of Enki who was a God of water. 35 is not a regular number or a divisor of 60.

Ninlil (45)

In Sumerian religion Ninlil is seen as the "Lady of the Wind", and she is the consort Goddess of Enlil, who was also a God of wind. 45 is a regular number but not a divisor of 60.

Antu (55)

Antu was the first consort of Anu, and the pair were the parents of the Anunnaki and the Utukki (evil deities). Antu was a dominant feature until she was replaced as consort by Ishtar, who may also be a daughter of Anu and Antu. 55 is also not a regular number or a divisor of 60.

35 and 55 are not regular numbers or divisors of 60, so they can't be included in the following chart. I don't know if this means that the deities for 35 and 55 are very important, or not that important.

If you look at all of the deities that have regular numbers at once, it is obvious that the most important deities usually get musical / regular number positions, and that regular numbers that are also divisors of 60 seem to be coveted positions for the younger deities. 60 has 12 divisors, if all of them had deities assigned, I have not found them all, hence the empty places at 1, 2, 3, 4 and 6 in these charts.

You can hear how 10, 25 and 45 fit with the other deities using the piano image below.

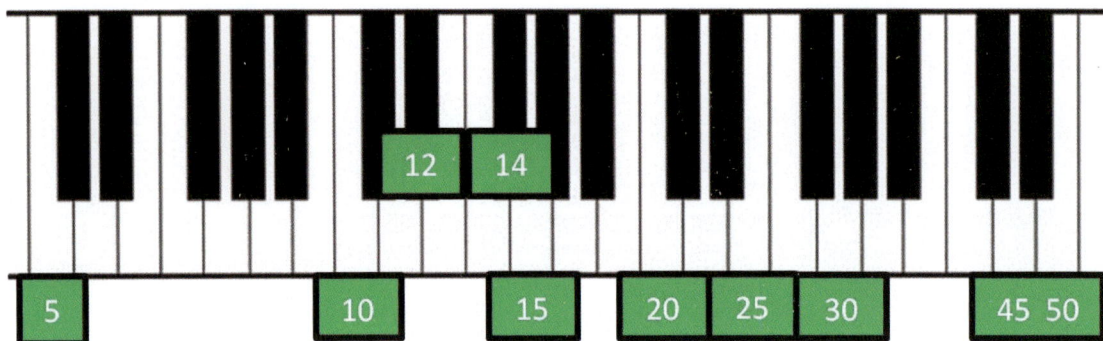

Deities	Regular numbers	Divisors of 60	Fraction of 60
	1	1	1 / 60
	2	2	1 / 30
	3	3	1 / 20
	4	4	1 / 15
Ninhursag	5	5	1 / 12
	6	6	1 / 10
	8		
	9		
Adad	10	10	1 / 6
Nabu	12	12	1 / 5
Ishtar	15	15	1 / 4
	16		
	18		
Shamash	20	20	1 / 3
	24		
Ningal	25		
	27		
Sin	30	30	1 / 2
	32		
	36		
Enki	40		
Ninlil	45		
	48		
Enlil	50		
	54		
Anu	60	60	1 / 1

Sum up

A lot is going on here, and as with all ancient things much of it can be understood in different ways…

What I think is most important though is that the numbers for the 7 highest deities make a major chord (stretched over a few octaves) that contains the building blocks for everything in this book.

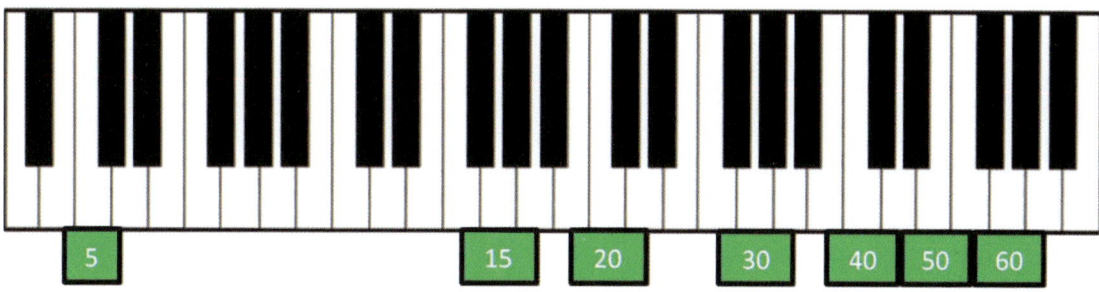

In it you can find the 2 – 3 – 5, 3 – 4 – 5 and 4 – 5 – 6 major chords laid out as 20 – 30 – 50, 30 – 40 – 50 and 40 – 50 – 60.

2 – 3 – 5 are the prime numbers that make all of the regular numbers and the 5 limit scale.

3 – 4 – 5 define the specific 3 polygons that the Platonic solids are made from.

4 – 5 – 6 include the hexagon found in truncated icosahedron based geodesic domes, Buckminsterfullerene molecules and soccer balls, snowflakes, bee hives, carbon nanotubes and many other places where strength is needed.

In conclusion

I think this is a fine point to end this book on. In my attempt to sum it up I'm going to give you a scenario, and it would be good if you could imagine it for yourself.

You're watching a horror movie. A hand holding a knife is approaching a shower curtain, behind which a lovely woman sings a tune from her heart. The soundtrack, however, builds up to a note that leaves you feeling tense, uneasy, scared, suspense-filled, anxious, nervous etc. It's those notes played at points in horror/thriller movies that are played using irregular numbers, or even totally irrational numbers. These are often made using bent pitches, metal objects and other non-musical sound generation methods.

Now imagine a different movie, where a lost dog finds its owners after months of searching through the mountains. The notes played in the music as the dog and owner hug, in slow motion, are usually made with the diatonic scale, which is based on 2-3-5; the major chord.

Typical atonal "horror / thriller" sounds have mathematics full of irregular or irrational numbers. These numbers are no fun to work with mathematically. These same numbers represent geometric shapes and amounts of shapes that are no good for building 3D objects. For example; the 7 based devil's interval is often used in this situation, and 7 sided polygons or 7 regular sided polygons also can't build anything nice and strong in 3D.

The "lost dog finds its owners" sounds, on the other hand, are based on mathematics that contains only regular numbers with the prime factors of 2, 3 and 5. These numbers are amazing to work with mathematically, making life much easier than it would be if you used irregular or irrational numbers. Regular number amounts of 2, 3 and 5 sided polygons are also what you use to make the Platonic and Archimedean solids, geodesic domes, pyramids, and many other strong and useful things.

At the end of the day, it is the prime numbers 2, 3 and 5 that connect everything together. They are the real reason why you find certain numbers like 144 and 432 with various amounts of 0's at the end in music scales, sacred geometry; the length of the day, the size of the Sun, various ancient texts and calendars, and the measures in many ancient monuments.

It may not seem profound but it is, because these 3 numbers also represent the most harmonious sounding chord in music, the major chord. This truly is the chord you would expect to hear in a movie at a moment of some great revelation.

I hope this book has been insightful for you, and I hope it has helped you to understand the very interesting connections between music, geometry and mathematics.

If you want to apply this knowledge in digital music production, you might like my other book "Mathemagical Music Production".

Facebook group to discuss these matters: "Music, Geometry and Mathematics".

Printed in Great Britain
by Amazon